Forecasting the Future

Exploring Evidence for Global Climate Change

NSTA Stock Number PB118X
SBAM Publication Number ED0003
C[4] Publication Number C4158

Library of Congress Catalog Card Number 96-67632

ISBN 0-87355-139-7

Printed in the U.S.A. by Graphic Communications, Inc.

The National Science Teachers Association is an organization of science education professionals and has as its purpose the stimulation, improvement, and coordination of science teaching and learning.

Forecasting the Future

Exploring Evidence for Global Climate Change

By

Education Department,
Stephen Birch Aquarium-Museum

in collaboration with

Center for Clouds, Chemistry and Climate
A National Science Foundation Science Technology Center

Scripps Institution of Oceanography, University of California, San Diego

Published by

National Science Teachers Association
1840 Wilson Boulevard
Arlington, VA 22201-3000

Table of Contents

Introduction

Acknowledgments...ii

Introduction..iii

Preface..v

Chapters

Clues to Past Climate..1

The Search for Long-Term Evidence..9

Current Global Climate Change...15

Water and Climate Change...27

Life and Climate Change..35

Activities

On the Move (Animal Biology)...49

A Fishy Tale (Animal Biology)..54

Air Cares (Chemistry)..58

It's Soup (Chemistry)..63

(Gonna Take a) Sedimental Journey (Geology)..67

Rocky Records (Geology)..71

Digging It (Geology)...78

Going, Going, Gone (Meteorology)...82

Making Weather: A Recipe Using Water (Meteorology)...86

It All Adds Up (Meteorology)...90

Flour Shower (Physics)...93

The Reasons for Seasons (Physics)..97

Plant Power (Plant Biology)...102

Leafing Through the Past (Plant Biology)..106

Extension Activities

Designing Science Lessons...111

Easy Extension Activities...114

Appendices

Earth Through Time..127

Glossary..134

Annotated Bibliography..138

Metric Conversion Table...148

Dedication

This work is dedicated to Donald Hilvert, Alex Hamolsky Hilvert, and Andres Alberto Hamolsky Hilvert, on whom the sun rises and sets. Thank you always, Esperanza, Sidney, Sharon, and George Hamolsky.

Acknowledgments

Forecasting the Future: Exploring Evidence for Global Climate Change is made possible through the generous support and dedication of more than one hundred teachers and scientists across the United States who read, reviewed, and tested the material. Scripps Institution of Oceanography and NSTA are profoundly grateful to these teachers and scientists, who contributed to the betterment of *Forecasting the Future* time and time again.

Forecasting the Future: Exploring Evidence for Global Climate Change was produced by the Stephen Birch Aquarium-Museum: Ned A. Smith (Executive Director), Ruth G. Shelly (Managing Director), Diane A. Baxter (Education Curator), Monica Carmela Hamolsky (Director of Teacher Education and Outreach Programs), Cheryl Trinidad (Administrative Assistant), and Steven D. Cook (Publications Manager).

V. Ramanathan, Director, H. Nguyen, Assistant Director, and Sharon Roth Franks, Education Liaison of the Center for Clouds, Chemistry and Climate, were particularly instrumental in support of this publication.

Many thanks for its donation of illustration slides go to the National Center for Atmospheric Research in Boulder, CO.

Forecasting the Future: Exploring Evidence for Global Climate Change is published by NSTA: Gerry Wheeler (Executive Director), Phyllis Marcuccio (Executive Director for Publications). NSTA Special Publications produced *Forecasting the Future*—Shirley Watt Ireton (Managing Editor), Douglas M. Messier (Project Editor), Chris Findlay (Associate Editor), Michelle Eugeni (Assistant Editor), and Christina Frasch (Editorial Assistant). Auras Design created the design and timeline. Steven D. Cook designed the cover. The book was printed by Graphic Communications, Inc.

Introduction

Welcome to *Forecasting the Future: Exploring Evidence for Global Climate Change*. This classroom curriculum and activities guide was developed by the Education Department of Stephen Birch Aquarium-Museum in collaboration with the Center for Clouds, Chemistry and Climate at Scripps Institution of Oceanography, University of California, San Diego. *Forecasting the Future* is published by the National Science Teachers Association.

We suggest you read through *Forecasting the Future* before beginning to teach from it. The publication's activities and approaches were developed to meet the skill levels and learning styles of a diverse student population. The narrative section, pages 1–48, provides background information on the subject. The narrative includes references in the form of icons to relevant activities in the second section of the book. If the text is difficult for students to read on their own, you might wish to interpret or adapt it to their interests and abilities.

The activities section, pages 49–110, provides 14 detailed exercises that illustrate ideas set out in the narrative. These hands-on activities represent various disciplines, including animal biology, chemistry, geology, meteorology, physics, and plant biology. They list objectives, estimates of duration, extended background information, introductory exercises, procedures, and useful discussion points. The activities are photocopy-ready and illustrated with procedural diagrams. They can be used in whatever order is best for individual programs.

A Teacher's Guide that provides additional information accompanies each activity. Answers to questions and pointers for success follow a general summary of the exercise.

The publication's third section, pages 111–126, includes an overview of scientific inquiry called Designing Science Lessons as well as more than 40 Easy Extension Activities. The Designing Science Lessons section will assist in creating independent learning opportunities. This knowledge can be applied to the Easy Extension Activities, which are brief summaries of additional activities designed to augment material found in the rest of the publication.

The final section, pages 127–148, includes a timeline called Earth Through Time, a Glossary, an Annotated Bibliography, and a Metric Conversion Table. Earth Through Time depicts interactive events in geological, biological, paleontological, and meteorological history over more than four billion years. The Glossary contains definitions of key words and concepts found throughout the book. The bibliography includes detailed descriptions of books, teacher guides, and Internet resources that relate to global change research. Contact and ordering information is provided. The Metric Conversion Table includes information about how to convert metric measurements found in the text to English measurements.

This is one in a series of ocean education resource guides developed and produced by the Stephen Birch Aquarium-Museum. The Aquarium-Museum's education programs offer workshops and symposia, resource materials for hands-on science activities, classroom computer software, pictorial slides, videotapes, and computer question-and-answer consultative services. They also provide teachers with opportunities to meet and interview practicing scientists, and to obtain in-class consultation and assistance from our staff. All can be used as part of the science curriculum.

To make suggestions on how to improve this publication, please contact: Monica Carmela Hamolsky, Stephen Birch Aquarium-Museum, SIO, UCSD, 9500 Gilman Drive, Department 0207, La Jolla, CA 92093-0207. We predict a bright forecast in your classroom through the use of this curriculum and activity guide, and we thank you for your interest in it.

Preface

How great is science's power to predict? How accurately are Earth's most vital signs being monitored? Images of an endangered Earth are helping to create public concern about impending climate change. Increased awareness of global environmental change is causing many individuals to alter their values and actions for the betterment of the world.

Yet prudent decisions rest on careful consideration of scientific evidence. Is the temperature of the Earth's atmosphere increasing? Are forecasts of a changing global climate warranted? If so, are they cause for concern, or do they represent normal change in a climate cycle of rising and falling temperatures? If global climate change is taking place, is it inexorable?

This curriculum and classroom activity guide considers evidence gathered in answer to these questions. It describes methods that biologists, chemists, geologists, meteorologists, and physicists use to gather and interpret their findings. Explanations given here reflect broad agreement among scientists about important aspects of global climate change and the possible effects, although not all scientists subscribe to the consensus. Diversity of opinion is natural and healthy in science, especially about a topic experiencing such rapid research advances as global climate change. For this reason, questions about future warming trends cannot be answered with finality. Above all, science is a continuing search for new information. It offers few absolute answers because its findings are constantly under study. With regard to global climate change, science's understanding is incomplete but increasing.

Forecasting the Future considers questions and issues of great importance to humankind. It also provides a picture of how science works to explain them.

1

Clues to Past Climate

Is climate change normal and cyclic?

What clues do scientists study to find out?

Seventy-five million years ago, the area we now know as Canada was considerably warmer. The climate across the entire Northern Hemisphere was at least 5.5°C warmer and a good deal more humid. How do we know that? Fossils show it—fossils found in Canada, Asia, and the northern United States; in mud and rocks; on land and in oceans; and in kilometers-deep, pitch-dark caves. For instance, fossils from Canada, Utah, Montana, and Michigan show that palm trees once grew much farther north than they do now. These warm-climate plant fossils give evidence that the Northern Hemisphere once had a much warmer climate than today. Although the intensity of warming probably varied from region to region, the climate was warmer and more humid almost everywhere in the world.

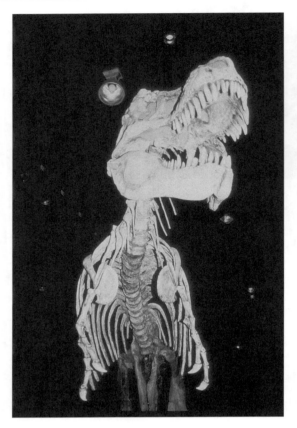

Fossils of trees and leaves offer clues to the composition of past atmospheres.

T. Rex and other fossils indicate past warm-climate conditions in now chilly northern regions.

Fossils of *Tyrannosaurus rex* have also been found, in both North America and Asia. (These two continents are now separated by the Bering Strait. To travel from one to the other, dinosaurs must have used a strand of land that is now underwater.) How could warm-climate dinosaurs survive so far north? For *T. rex* to have ranged over North America and Asia for millions of years, winters must have been much milder than they are now. Today's reptiles—alligators, for example—cannot live in cold climates. In the United States, alligators' range extends only a little north of the Gulf Coast. But fossils of alligators from five million years ago have been found as far north as Canada, another indication of a once-warmer climate.

Now extinct, ammonites were shelled animals that lived in the sea. Surprisingly, their shells have been found in the center of our continent, far from present coasts. This suggests that seawater once extended across North America, and that this water was warm enough for the ammonites to survive. Some scientists think that a huge sea once covered a large part of today's continents, and that this

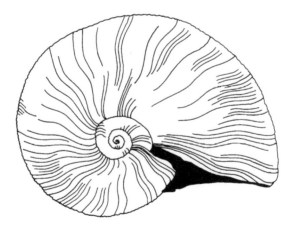

Marine fossils found on land indicate that the land was once covered by oceans.

sea helped to keep Earth's climate mild. Why do they think so? How could more extensive oceans affect climate change? What other factors might also have warmed the planet?

What Made the World Warmer?

Scientists have developed at least two prevailing theories. First, partly because the Earth had more extensive oceans, the atmosphere may have contained more of the gases that trap heat: water vapor (H_2O) and carbon dioxide (CO_2). Volcanoes also contributed these gases, along with sulfur-rich gases (*e.g.*, SO_2) that form heat-trapping particles. Second, Earth might have been warmed by the Sun more than it is today.

The Atmosphere May Have Contained More Heat-Trapping Gases

Scientists believe much more water once lay over the world. It is possible that 85 percent of Earth was covered by water, in comparison to approximately 70 percent today. This may have increased the amount of water evaporated into the atmosphere. If so, a part of this water vapor condensed to form clouds, and changes in their number and distribution would have resulted.

More clouds could have trapped heat rising from Earth's surface. Furthermore, increased cloudiness may have blocked the sunlight that plants require for photosynthesis. Plants prevented from growing would have removed far less CO_2 from the atmosphere. Therefore, CO_2 concentrations in the atmosphere would have increased.

More volcanic activity in the past spewed great amounts of water vapor and CO_2 into the atmosphere. Volcanoes also emit sulfur-rich gases, which form particles that can trap heat.

Meteorologists and other scientists study fine-grained particles, droplets, and gases thrust up into the atmosphere by volcanic eruptions in Mt. Fuego, Guatemala, among other sites. Volcanic eruptions such as this one emit heat-trapping gases into the atmosphere.

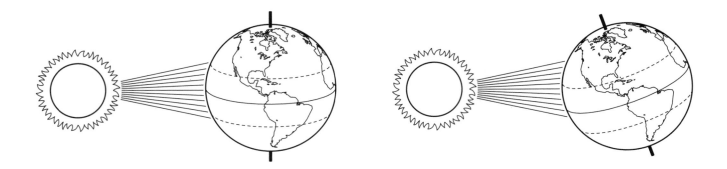

Changes in the tilt of Earth's axis altered the distribution of solar energy received by our planet.

The Earth might have absorbed and spread more of the Sun's warming rays

The Earth's tilt and orbit around the Sun change periodically. These shifts may have increased the amount of sunlight reaching part or all of the planet. As a result, temperatures could have risen.

PHYSICS
The Reasons for Seasons

As temperatures rose, there was gradually less ice and snow to reflect sunlight back into space. The exposed dark surface of ice-free ground and waters would have soaked up more and more heat.

If more water covered the Earth, currents could have flowed unimpeded by land barriers, distributing heat more evenly around the globe. On a much longer time scale, another factor facilitating heat transport may have been the placement of land. Continents were positioned differently in the past, and currents may have flowed more freely.

AND NOW FOR A SURPRISING SWITCH!

More Recently, It Was a Colder World

Some years ago, fishermen working the Atlantic Coast off New Jersey hauled up the giant fossil tooth of a mastodon. So much fossil evidence of land mammals has been found offshore that most scientists believe mastodons once walked on land extending far into what is now the Atlantic Ocean. This means that sea levels must have been far lower at one time. What could cause sea levels to change? In times of cold climate, ocean levels recede as water freezes into ice. So much water was locked into ice about 20,000 years ago that less liquid was left in the oceans and more dry land was revealed.

Many scientists believe that ice covered nearly one-third of the planet's land surface many thousands of years ago. (Today it covers only one-ninth.) Core samples, fossils, and other evidence suggest that one immense ice sheet, 3.2-km thick in places, buried most of Canada and reached as far south as northern Illinois. Other massive ice sheets are believed to have covered Greenland, much of Northern Europe, and most of Asia.

Fossil records indicate that over the past million years, huge ice sheets advanced and retreated many times. Very cold periods have come and gone. Most scientists believe that at the peak of the last ice advance, 20,000 years ago, the global temperature average was about 5°C lower than it is today. Much of our northern continent was buried under colossal sheets of ice. During this time, animals such as reindeer, wolverines, and lemmings lived in areas now much too warm for them. Today, musk oxen are found only in the Arctic. Once, however, their range extended all the way to Mexico. Today, walruses live only in frigid waters, but fossils show that walruses once swam as far south as the coast of Virginia. For walruses to have survived there, the climate must have been considerably colder than it is today.

What Made the World Colder?

Again, at least two theories explain why the Earth may have been colder during earlier times.

Earth May have Absorbed and Spread Less of the Sun's Warming Rays

Earth's tilt and orbit around the Sun change periodically. At some points in time, these shifts decreased the amount of sunlight reaching part or all of the planet. Temperatures dropped, and snow gradually accumulated. Glaciers grew larger and advanced across the land.

As temperatures fell, ice and snow covered more of the Earth's surface. Light-colored surfaces absorb less heat than dark ones, so less of the Sun's warmth was retained on Earth. In fact, rather than retaining heat, increased ice and snow would have reflected more sunlight back into space.

Earth's waters chilled as well. More water froze into ice, and frozen waters cannot transport and distribute warmth. Furthermore, like land cov-

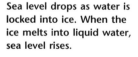

Sea level drops as water is locked into ice. When the ice melts into liquid water, sea level rises.

Walruses, such as these photographed in the Northwest Territories of Canada, live in cold climates. Fossil evidence shows that walruses once lived as far south as Virginia, indicating that climate in the region was once much colder.

ered with snow and ice, water in a frozen state can absorb and retain little heat. It too reflected away the Sun's heat.

The Atmosphere May Have Contained Fewer Gases to Trap Heat

If less liquid water lay over the land, the amount of water evaporated into the atmosphere may have decreased. A lessening of the cloud cover would have resulted, so less heat from the surface might have been trapped.

If there were fewer or smaller clouds, more sunlight reached and warmed the Earth. Increased sunlight may have benefited terrestrial and marine plants. If so, these plants would have taken up greater quantities of CO_2.

Atmospheric particles from volcanoes' sulfur form a shield-like haze that reflects sunlight. This cools the Earth's surface. These particles also absorb heat rising from the Earth's surface, thereby warming the atmosphere, but their cooling effect dominates.

Periodic evidence of receding waters (due, in this case, to drought) may be observed at Shasta Lake in California.

Why Do Climate Cycles Change Direction?

By now, careful readers have undoubtedly noticed something surprising. Some factors in climate shifts—factors such as volcanic activity, changes in the Earth's orbit, or changes in cloud cover—might have more than one effect. They may even have opposite effects. For instance, lessened cloud cover is one factor described here as having two opposite impacts on temperature. On the one hand, sparser clouds trap less of the heat that rises from the Earth. The overall effect is a cooling one. On the other hand, sparser clouds permit more sunlight to reach the Earth, thereby warming it. Which effect of lessened cloud cover had more impact in the past: the cooling effect or the warming effect? Did one outweigh the other? Is it possible that the two cancelled each other out? The short answer is that we don't know.

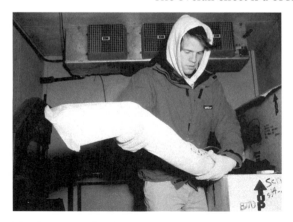

Scientists can analyze ice and air trapped in ice to build a record of past temperatures.

Why is the influence of cloud cover on climate imperfectly understood? Why are scientists unable to say with complete certainty what caused climate to shift from warmer to colder periods—and back again? The reason is that no single factor can account for climate change. In fact, any one change generally causes another, which causes a third, and so forth. Ultimately, for climate change to occur, a series of events must take place and stay in place for many years.

METEOROLOGY
Going, Going, Gone

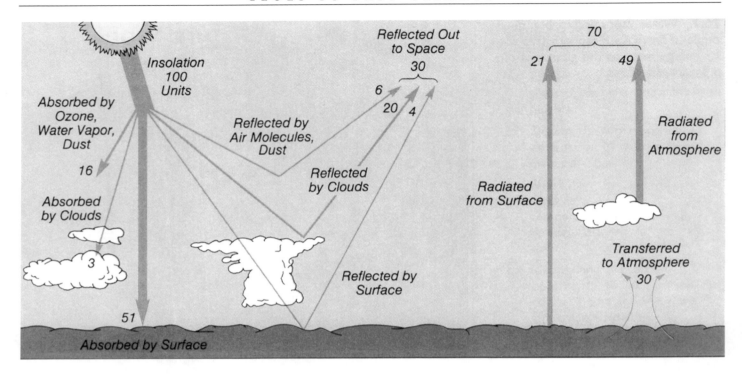

This diagram shows what happens to 100 units of sunlight entering the Earth's atmosphere. Only 51 units are absorbed by the surface. Notice that the clouds have dual functions; they both reflect and absorb sunlight.

For example: if planetary tilt causes temperatures to rise on part of the planet, some of the world's waters warm. Warmer waters evaporate more quickly, so they might create more cloud cover. More clouds could lead to less light penetration, poorer plant growth, more atmospheric carbon dioxide, more trapped heat, and even warmer temperatures.

In this scenario, we find ourselves back where we began. Warming led to more warming. This is an example of a *positive feedback loop*. It is positive because the steps in the cycle resulted in more of the original trend. If this were the only direction that climate cycles take, the atmosphere would become warmer and warmer, and the oceans would eventually boil away. Above approximately 110°C, life would end.

Fortunately, there are *negative*, or counteracting, feedback loops. Let's begin as we did earlier, with the evaporation of warm waters. In a second scenario, more evaporation might lead to more clouds, less light penetration, cooler temperatures and less evaporation. This is an example of steps in a cycle *reversing* the initial trend.

The interaction of feedback cycles is such that it is not always possible to know whether climate change is the cause, the result, or a side effect of other changes. Furthermore, many changes can contribute either to warming or to cooling temperatures. As we shall see, changes in climate depend on what occurs beforehand, simultaneously, and afterward.

2

The Search for Long -Term Evidence

How do scientists know what the climate was many thousands of years ago?

Where do they search for clues?

For a reliable picture of Earth's past climate, scientists consider evidence from many sources. Seeking climate clues, they probe centuries-old trees, ancient caves, ocean and lake floors, and glaciers. Putting together the information they find, they construct a continuous record of past temperatures, rainfall, vegetation, and levels of atmospheric gases. This historic record dates back hundreds to millions of years.

N S T A

Climate records are preserved in tree rings. The thickness of each annual ring reflects growing conditions, including climate, during ring formation.

Tree Rings

How do scientists read tree rings? It isn't necessary to cut down a tree to study its rings; researchers can drill horizontal cores from the tree, and study the core to find evidence of past climate changes. (Coring doesn't harm the tree. Woodpeckers drill holes in trees all the time without damaging them.) Thin plugs of wood removed from the tree show sections of its annual rings.

A growing tree adds a new ring of wood to its trunk every year, so *counting* the rings tells the tree's age. But *measuring* the thickness of the rings gives even more information. Trees grow well during years when weather conditions are favorable. They produce wider rings than in years when moisture and temperature conditions are less favorable. So the width or narrowness of each ring tells of climate conditions each year, every year, often over hundreds of years. Some changes in ring width are not obvious to the naked eye, so scientists use magnifying lenses and microscopes to measure them. In addition to trees living today, scientists study the petrified remains of trees that lived hundreds, thousands, and even millions of years ago. These remnants provide valuable evidence of climate conditions from the Earth's distant past.

PLANT BIOLOGY
Leafing Through the Past

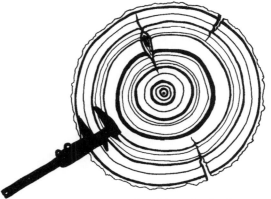

Scientists measure the thickness of tree rings with an instrument called a micrometer.

Stalactites and Stalagmites

Limestone spires form slowly, over many thousands of years, as water drips through limestone caves onto the ground below. Stalactites hang like stone icicles from cave ceilings. Stalagmites rise from the floor, tapering upward. Geologists refer to these two types of spires as "speleothems." They accumu-

late when the calcite in limestone—dissolved in water seeping through the cave—re-solidifies within the cave. This buildup of calcite is similar to the buildup of mineral deposits on water pipes and shower doors.

Scientists use various oxygen-sampling devices to study past temperatures and climates.

In cross-section, stalactites and stalagmites typically have a banded structure resembling growth rings in a tree trunk. As in trees, the bands represent periods of expansion. Unlike tree rings, they don't form every year. Speleothems grow only when the weather is warm and there's ample rain; the thickest layers indicate warmest, wettest conditions. When geologists find some thick layers and some thin ones in a stony spire, this is evidence that the climate varied in the past. However, in contrast to tree rings, not only the thickness but the very composition of a stalactite or stalagmite tells a story.

GEOLOGY
Rocky Records

When water drips down through the roof of a cave, it is sometimes enclosed by the minerals it carries. Water drops shielded by minerals do not evaporate. In fact, they often stay intact over the course of centuries. Scientists study the water and the mineral deposits for clues about ancient climates. To estimate past temperatures, for instance, scientists compare the amounts of two kinds, or isotopes, of oxygen present in the mineral calcite. To date those temperatures, scientists compare the relative presence of uranium and thorium atoms in the minerals. The older the stalagmite section, the more thorium its water will contain, because uranium atoms decay into thorium.

Deep-Sea and Lake Sediment

Sediment cores often look like mud, but to researchers that mud can be true treasure. The mud contains the remains of plants, animals, and particles (such as dust and clay) that have accumulated over eons at the bottom of oceans and lakes. Scientists extract long cylinders, or cores, of these sediments, and analyze the contents for clues about the climate at the time the materials were deposited. The oldest sediment is usually at the bottom of the core. The material becomes younger and younger toward the top of the core because new sediments sink through the water and settle on top of older ones. In some cases, however, earthquakes and other geological activities disturb this layering process, pushing the older sediment toward the top. For this reason, care must be taken in examining cores.

GEOLOGY
Sedimental Journey

A single deep-sea core can show sediment buildup over millions of years. Close to land, where runoff washes large amounts of material into the ocean, the rate of sediment accumulation on the seafloor can be several meters per thousand years. Farther from shore, accumulation rates are much lower: only a few meters per million years.

The makeup and the amount of material that settles to the seafloor depends on a number of factors. These include: (1) the depth and chemistry of the water; (2) the distance of the site from land; (3) regional currents; (4) the climate; and (5) seasonal weather conditions. Because all these conditions change over time, the composition of mud changes along the length of a core. Scientists describe and date these changes, using this information to help develop a history of past climate. For example, sediments' color often reflects the amount of oxygen present over time. This tells us something about the vigor of ocean circulation, positioning of continental land masses, global sea level, and climate.

Even the grain size of sediments reveals much about conditions in the past. Typically, near-shore sediments are deposited where currents are strongest. They tend to be coarse, while deep-sea sediments are generally finer-grained. Given this, changes in grain size along the length of a core suggest changes in distance between the sampling site and the shore. Accumulation patterns often shift due to fluctuating sea levels, and these fluctuations can be caused by changes in climate. (It is also true, however, that differences in sedimentation sometimes have nothing to do with temperature change. Sand accumulation might simply result from plate tectonic rearrangement of land masses.)

Valuable information about climate conditions long ago is also locked

Following their first appearance in the Cambrian period, many different species of sea-dwelling ammonites lived and died out, making them useful fossils for dating objects and events.

into tiny animals' shells or spines. When a marine animal dies, its soft tissue becomes part of the ocean's floating food mix. It is generally eaten. If the tissue isn't eaten, it decays. Animals' hard parts, however, are generally neither edible nor quick to decompose, so they ultimately settle to the sea floor to become part of the sediment record.

How do these preserved parts tell scientists about past climate? Biologists can compare the shells in the core with similar, modern shells. Most modern organisms' geographical range is known. Scientists can differentiate between animals adapted to warm waters and those requiring colder conditions. Some marine organisms live on rocky shores, while others prefer the sandy sea bottom or sunlit surface waters. Knowing this, biologists can infer that past life forms probably lived in locations having certain temperature ranges, just as their modern relatives do.

Other evidence of past climate conditions is provided by shells of tiny animals called foraminifera—forams for short. While they are living, forams use two types, or isotopes, of oxygen to build their shells. If the waters are warm, more of one type is used than the other. When the forams die, the shells drift to the ocean floor and are preserved in the sediment. Measuring the relative levels of oxygen isotopes in the shells provides another way of gauging temperatures long ago.

Scientists take and analyze core samples from lake-bottoms to determine the levels of plant remains, silt, and clay. Another indication of climate comes from the proportion of organic (*e.g.*, plant remains) and inorganic (*e.g.*, clay particles) sediments found in these samples. Different plants thrive at different levels of temperature and precipitation. By identifying leaves, seeds, pollen, and bits of wood in lake cores, scientists can estimate local climate conditions at the time those plants were growing. If little organic material is present, conditions were not very favorable for plant growth. In

ANIMAL BIOLOGY
On the Move

ANIMAL BIOLOGY
A Fishy Tale

contrast, lake cores containing much organic matter, like pollen, suggest that vegetation was denser because climate and other growing conditions were favorable. Alternatively, when climate conditions hamper plant growth, soil is more susceptible to erosion. More soil and clay are washed into lakes, later to be found in cores.

Glacier Ice

Another way scientists gather evidence of past climate change is by studying ice cores extracted from glaciers in Antarctica and Greenland. From these cores, scientists have identified climate changes over the last 100,000 years. That's not as long as the million-year records from deep-sea sediment cores, but ice core information is especially valuable because it shows changes that occurred over time periods as small as one year. Polar ice forms up to 50 times faster than deep-sea sediments, so it gives much better time resolution than geologists can obtain from sediments.

Glaciers are indicators of long-term temperature changes in that their movement etches evidence of climatic change on the land.

Information in ice cores comes from three sources: (1) from the water making up the ice itself; (2) from gases such as oxygen, carbon dioxide and methane trapped in air bubbles; and (3) from impurities such as dust, volcanic ash, salts and industrial emissions found in the ice. By analyzing the ice and the air trapped in it, scientists can build a yearly record of past temperature changes that goes back thousands of years. With respect to impurities, differences in amount might tell of drier, dustier or windier times. These can be linked to periods of great cold. For instance, there is twenty times more dust and ash in ice that formed 18,000 years ago, during the coldest part of the last glacial period, than in ice formed within the last 10,000 years. This suggests that during colder periods the land surface was on average drier and relatively easily swirled out of place by winds.

3

Current Global Climate Change

Is global warming occurring?

*How do scientists take
the planet's temperature to find out?*

Climatologists have observed a slight but fairly
steady rise in temperatures since the 1880s,
when reliable temperature records first became
available worldwide. What does this rise tell
us? To know if increasing temperatures signal
global climate change, we must first be sure
that the temperature measurements represent
the entire planet—that they really are *global*.

A compilation of three different photographs taken from spacecraft, this view of the blue planet Earth shows heavy cloud cover in the Southern Hemisphere.

METEOROLOGY
It All Adds Up

Meteorologists try to make sure measurements cover the entire globe by dividing it into an enormous, imaginary grid and collecting temperatures from weather stations in each square. To keep track of temperatures in each square, scientists rely on specially-fitted satellites and weather stations on the ground and at sea. Averaging all the measurements produced by these stations yields a global average temperature. Plotting these averages on a graph enables meteorologists to observe trends in rising or falling temperatures. Other indicators of climate are also collected: levels of humidity and rainfall, the distribution and depth of snow and ice, and changes in winds and currents.

To know if global climate change is occurring, we must also be sure that we are correctly measuring *climate*. For one thing, we must be sure to avoid confusing climate and weather. *Weather* refers to atmospheric conditions on a short time scale: days or weeks. If it's sunny and hot today, and rainy and cooler tomorrow, there's been a change in weather. In contrast, climate refers to conditions over a longer time scale: years, decades, centuries or millennia. *Global climate change,* therefore, refers to a worldwide shift in atmospheric conditions extending over years, decades, centuries or millennia. By this definition, we have over the last hundred years experienced slight global warming.

Will this warming persist? Will it intensify? If so, for how long? Climate responds to so many forces that it is hard to say. There are astronomical influences on climate, such as changes in Earth's orbit around the Sun or changes in the distribution of solar energy sweeping Earth's surface. Geologi-

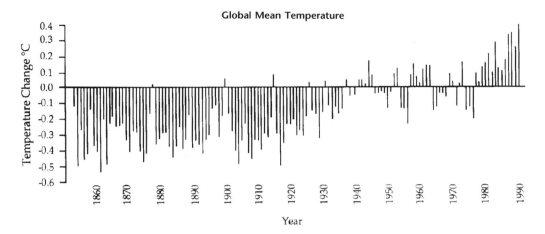

Global Mean Temperature

Deviations from a global temperature of 15°C for the period from 1854 to 1990. Data from the National Oceanic and Atmospheric Administration (NOAA).

cal forces also influence global climate: plate tectonics carry land masses to warmer or cooler latitudes; volcanic activity intensifies or drops; sea levels change. Meteorological changes in cloud cover and winds contribute to climate. Chemical processes in the oceans and atmosphere come into play, as do alterations in the life cycles of plants and animals.

All these contributions to global climate change are natural. In other words, the changes occur irrespective of the presence of human beings on Earth. We can say with certainty that astronomical, geological, and meteorological processes of the distant past had nothing to do with human beings. Humans have gained greater ability to alter our environment, however. There is increasing evidence that our activities can affect global climate. For example, industrial pollutants produce an airborne haze that absorbs heat, warming the atmosphere around it. We are also adding gases to the atmosphere that appear to have warmed the planet over the last one hundred years. These gases are called "greenhouse" gases.

The Greenhouse Effect

A greenhouse is a building whose walls and ceilings are made of clear glass or plastic. It is called a greenhouse because gardeners can grow plants such as ferns, flowers, vegetables, and herbs in it all year, even during the winter when temperatures outside the greenhouse are freezing.

A greenhouse makes it possible to grow plants in winter because its panes are transparent to the Sun's warmth. The panes let light in: sunlight passes through greenhouse glass and warms the ground inside. When the ground has warmed sufficiently, heat begins to rise from it and warms the air. Ceiling and walls prevent the warm air from escaping, so the greenhouse retains the heat.

How the atmosphere traps heat

The atmosphere might appear to be an enormous zone of empty space, but it isn't. Though they aren't visible to the naked eye, molecules of many gases are moving about in the atmosphere. The air we breathe, for instance, is rich in nitrogen (N_2, a molecule composed of two atoms of nitrogen), and oxygen (O_2, a molecule made of two atoms of oxygen). These molecules allow visible light, one type of solar energy, to pass through them. They also allow other, invisible light waves from the Sun to pass through. Among these invisible light waves are those produced just beyond the red end of the light spectrum. This type of solar energy is called infrared radiation. Although it cannot be

CHEMISTRY
Air Cares

seen by human eyes, infrared radiation can be felt as heat.

The Sun's rays are made up of waves, or electromagnetic vibrations, that move at different speeds. Sunlight is a mixture of these waves. Many gas molecules, especially those made up of at least three atoms—*e.g.*, carbon dioxide (CO_2); methane (CH_4); nitrous oxide (N_2O); water vapor (H_2O); and ozone (O_3)—

PHYSICS
Flour Shower

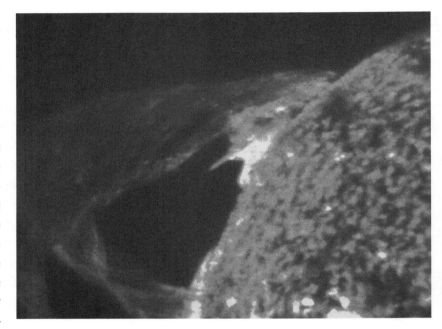

Heated gases erupting from the surface of the Sun escape into space.

allow some but not all light waves to pass through them. In fact, triatomic (three-atom) molecules allow most visible light and some wavelengths of infrared radiation to reach the ground. However, almost all outgoing infrared waves, re-radiated and rising from the Earth's surface, are blocked. As infrared energy ascends into the atmosphere, triatomic molecules absorb it and return it earthward. Warm temperatures result. Like a blanket or coat that prevents body heat from dissipating into cold air, atmospheric gases made up of triatomic molecules hold heat close to our planet.

Life Depends on the Greenhouse Effect

You may have gathered from media coverage that the greenhouse effect is dangerous. That is simply not true! The greenhouse effect is essential to life on Earth. It maintains the average surface temperature on our planet at a comfortable 15.5°C. If there were no greenhouse gases in the atmosphere, most of the heat radiated from the Earth's surface would rise away from us into outer space. The planet's temperature would be well below freezing, far too cold for most life.

Planetary temperatures in our solar system depend largely on a planet's distance from the Sun, but temperature variations are also due to differences in their atmospheres. Mars, which is further away from the Sun than Earth, has a very thin atmosphere and weak greenhouse effect. Its surface is much colder than ours; in fact, it is frozen. Venus, which is closer to the Sun, has an intense greenhouse effect. Its atmosphere is so thick with CO_2 and other heat-trapping gases that its surface is hot enough to melt lead.

Earth's atmosphere has always contained greenhouse gases. These gases have created a natural warming effect that sustains life on Earth. Given this, current concern about the greenhouse effect does not center on the atmosphere's beneficial effects. Instead, we worry about enlarging the greenhouse effect. Human beings are adding heat-trapping gases to the atmosphere at a rate that could cause this planet to overheat. By increasing naturally occurring heat-trapping gases and adding new ones, we may be intensifying the already-existing greenhouse effect beyond our capacity to correct or reverse it.

Global warming is not caused by a heater that can be switched off if the Earth warms too much or too quickly. On the contrary, some greenhouse gases linger in the atmosphere, trapping heat for many decades. What makes greenhouse gases so effective at trapping heat? The gases' effectiveness is determined by their **configuration**, **quantity**, and **residence time**.

Configuration. Atomic arrangement is an important determinant of a gas's heat-trapping activity. The atoms making up a molecule of water vapor, for instance, are arranged differently from those in a molecule of carbon dioxide. There are differences in the distance between atoms, and there are also differences in the amount of energy holding the atoms together. Physical properties such as these determine the radiation wavelengths that gas molecules absorb, and the amount of radiative heat absorbed is a key aspect of greenhouse gas effectiveness.

Quantity. Two crucial contributors to a gas's heat-trapping activity are their concentration and the rates at which those concentrations are changing. For example, the amount of water vapor in the atmosphere is large but relatively constant. In other words, its rate of concentration change is slow. In the case of carbon dioxide, levels are comparatively low but rising. Its rate of change is greater than water vapor's, though not as high as those of other greenhouse gases. Methane is a gas whose current atmospheric concentration is low relative to water vapor and carbon dioxide. Methane is present in very small amounts, but is being added to the atmosphere much more quickly than in previous decades.

Residence Time. Some gases trap limited amounts of heat because their residence time in the atmosphere is comparatively short. Many gases undergo chemical changes when they enter the atmosphere. Some are dispersed or destroyed, and some are transformed into states that do not trap heat. These

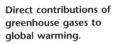

Direct contributions of greenhouse gases to global warming.

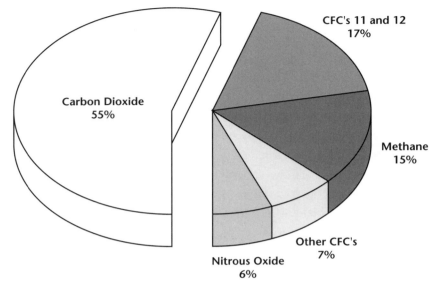

gases can do only limited damage because they are present in the atmosphere for only a short time. Other gases, however, resist reactions that might remove them from the atmosphere. Such gases are comparatively inert. In other words, they continue to trap heat in the atmosphere for a long time.

The above chart shows how much each of the gases we add to the atmosphere contributes to overall human-caused warming. It is important to understand that these gases are proportionally minute in an atmosphere largely made up of nitrogen and oxygen. For example, carbon dioxide constitutes only 350 parts per million of the atmosphere, just three-and-one-half hundredths of one percent. Atmospheric concentrations of methane, nitrous oxide, and CFCs are even lower. It is a sobering indication of carbon dioxide's heat-trapping effectiveness that it accounts for over half of the predicted increase in the greenhouse effect despite being present in such tiny amounts.

What will happen if carbon dioxide concentrations continue to increase? Only predictions are available, but one thing is certain: reducing greenhouse gas emissions now will minimize unwanted and potentially harmful effects in the future.

Greenhouse Gas 1: CO_2/CARBON DIOXIDE

We burn coal, oil, and natural gas to heat our homes, run our cars, light our streets, and power our factories. These fuels are called "fossil" fuels because they are the remains of plants that lived, died, and decayed millions of years ago. While living, plants depend on carbon dioxide in the air around them. In fact, they use energy from sunlight to transform the gas into carbohydrates (their food) and oxygen (a by-product). Scientists use an equation to tell this story:

$$CO_2 + H_2O \xrightarrow{\text{SUNLIGHT}} CH_2O + O_2$$

Industrial facilities–such as power plants, smelters, and refineries–are the source of a number of greenhouse gases, including CO_2 sulfur, and chlorofluorocarbons (CFCs). Scientists are concerned that these emissions will increase the atmosphere's ability to retain heat, thus raising global temperature.

PLANT BIOLOGY
Plant Power

This means: CO_2 (carbon dioxide) and H_2O (water) are changed by the Sun's energy to CH_2O (a carbohydrate building block) and O_2 (oxygen). Plants consume the carbohydrates, using them to make new tissue. (Animals, including human beings, consume the oxygen; we breathe it.) Burning both living and dead plants releases their carbon. Depending on the temperature of the heat source, the carbon is given off either in the form of particles or as a gas. When given off as a gas, the carbon combines with atmospheric oxygen to form CO_2.

Burning one pound of coal generates the electricity to light a 100-watt bulb for ten hours. For every pound of coal, nearly three pounds of CO_2 go into the atmosphere. Can you calculate how much CO_2 is generated in lighting your home or classroom in a day? A month? A year?

Consider another fossil fuel, gasoline, which is made from oil. Burning one gallon of gasoline generates approximately 20 pounds of CO_2. There are 500 million (5×10^8) gasoline-consuming motor vehicles in the world today. If present trends continue, the number of cars on Earth will double in the next thirty years to one billion (1×10^9). This means the potential for increasing CO_2 concentrations in the atmosphere is very high. To understand a single family's contribution to this increase, you can calculate how much CO_2 is added to the atmosphere by operating your family car over one year. For example, if your family uses 10 gallons of gasoline each week (or 520 gallons each year), 10,400 pounds of CO_2 are produced in one year from the operation of your car alone. To determine the total amount of CO_2 generated by automobiles worldwide, multiply this number by 5×10^8.

Greenhouse Gas 2: CH$_4$/METHANE

Methane traps heat twenty times more effectively than carbon dioxide. This means that each molecule of methane traps as much heat as twenty molecules of CO$_2$. Although currently causing less warming than CO$_2$, the increase of methane is greater than the increase of carbon dioxide. So not only is methane better at trapping heat, but humans add it to the atmosphere much faster than we add CO$_2$.

Where does this gas come from? Methane is made by bacteria that thrive in wet soil, so it is released from garbage landfills and open dumps. It also leaks out of the ground as coal, oil, and natural gas are mined. (Instead of allowing it to escape, we could collect methane and use it as fuel. Methane power plants do exist; one is located near a landfill site in Corvallis, Oregon.) And then there's rice. Rice, the world's most important grain crop, feeds one-third of its people. Over the last 45 years, farmland used for rice has doubled, and most rice grows in flooded fields. It is from this waterlogged soil that much methane is released.

Livestock are another source of methane. Every time cattle, sheep, and goats burp, methane is released. Bacteria in these animals' intestines break down the food they eat, converting some of it to methane gas. This is also true of other cud-chewing animals such as camels, buffalo, and deer. A cow, for instance, can belch almost one-tenth of a kilogram of methane a day, and there are now 1.3 billion cattle, each burping several times a minute. (Go ahead and laugh.) The problem is that increased demand for meat and dairy products to feed the world's rapidly-growing population has led to a doubling of the number of cattle in the last ten years, and the methane produced has a residence time of ten years. In other words, methane stays suspended in the atmosphere for ten years before it changes form. Even once

Production of methane is increasing in rice paddies, especially in those which are fertilized.

altered to a state that no longer actively traps heat, methane continues to cause warming. Chemical reactions removing methane from the atmosphere yield another greenhouse gas as a by-product: CO_2.

Greenhouse Gas 3: N_2O/NITROUS OXIDE

You or someone you know has probably encountered nitrous oxide in a visit to the dentist. When used as a drug, nitrous oxide blunts the sensation of pain. In this form, it is an anesthetic called laughing gas. However, the effects of this gas in the atmosphere are no laughing matter. N_2O traps heat 200 times more effectively than CO_2, so its atmospheric concentrations are cause for concern.

Nitrous oxide occurs naturally. Like methane, it is made by bacteria in soil. Human activities also generate N_2O, so concentrations of this gas are rising above natural levels. We add billions of tons of N_2O to the atmosphere each year, mainly through the use of nitrogen-based fertilizers, the disposal of human and animal wastes, and automobile exhaust. Most nitrogen-based fertilizers are used in growing crops that feed the world's ever-expanding population. Because these fertilizers speed and increase crop yields, their use has doubled in the past 15 years. An unwanted side effect of their decomposition in soil, however, is that nitrous oxide is released into the air. Also adding to the problem of rising atmospheric nitrous

Gaseous emissions from small-scale rooftop wetlands are studied to refine scientists' understanding of living organisms and their contribution to greenhouse gas accumulation.

Livestock (*e.g.*, cattle, sheep, goats, buffalo, and camels) produce methane in their digestive systems, contributing this heat-trapping gas to the atmosphere.

Monitoring instruments collect gases emitted by bogs and other wetlands in Canada. Sensitive as it is to environmental change, thawing tundra releases much methane and other heat-trapping gases into the atmosphere.

oxide concentrations are emissions of N_2O from sewage treatment plants and motor vehicle engines.

Greenhouse Gas 4: CFCs/CHLOROFLUOROCARBONS

Unlike nitrous oxide, methane, and carbon dioxide, chlorofluorocarbons do not occur in nature. They are human-created molecules used in industry—for air conditioning, refrigeration, electronics, packaging, and foams. When these molecules leak into the air, they are very effective at trapping heat. When they drift up through the atmosphere, they also destroy molecules in part of the upper atmosphere called the ozone layer. This is a problem because the ozone layer is Earth's essential shield against ultraviolet radiation. The ozone layer protects all living things by screening out this dangerous solar radiation.

Because CFCs are now known to break down the ozone layer, many countries have agreed to phase out their production. That's the good news. Unfortunately, some of the new substitutes for CFCs still trap heat. Furthermore, they have long lifetimes in the atmosphere. So even though they don't destroy ozone, they still contribute to global warming. That's the bad news. Fortunately, environmentally-safe alternatives to CFCs do exist. Although they are not yet used by many manufacturers, water- and helium-based technologies might some day replace CFCs.

Greenhouse Gas 5: H₂O/WATER VAPOR

What surrounds us at most times and in most places, yet is invisible? What is unseen until it transforms into tiny droplets or ice crystals? The answer

Scientists analyze biological processes such as photosynthesis and transpiration in the field by means of portable instruments that measure water quality and dissolved gases.

CHEMISTRY

Air Cares

is "water vapor," which is suspended in the air all around us but cannot be seen until it condenses into liquid water or freezes into ice. Unlike most greenhouse gases, which are present in such small amounts they are almost undetectable, water vapor can reach extremely high concentrations. Because of its substantial presence in the atmosphere, water vapor absorbs a great deal of heat radiated from the Earth's surface. Water vapor also acts as an amplifier. Increases in temperature due to any cause—carbon dioxide rise or orbital shifts, for example—intensify evaporation of water. This additional water vapor causes an even greater rise in atmospheric temperature.

But water vapor's activity as a heat–trapping and heat–intensifying greenhouse gas is not its only role in global climate change. This is because water

on Earth exists in three states: as a gas called water vapor, as a liquid called water, and as a solid called ice. If cooled, water vapor becomes liquid water. If cooled further, liquid water locks into ice. In those three forms, water plays a starring role in regulating the world's temperature.

Will water vapor contribute to global climate change? If so, how much of its influence will be due to water vapor's relation to the world's liquid waters and to its huge tracts of ice and snow? What part will be played by all the world's waters in our forecast of the future?

Moisture in updrafts—warm, rapidly rising air—condenses into raindrops or hail, falling to the ground when the updrafts can no longer carry the weight of the precipitation.

4

Water and Climate Change

What is water?

How do scientists study its effects on global climate?

Water is a substance made up of three atoms: two atoms of hydrogen and one atom of oxygen. These three atoms are held together by electrical forces, or bonds, in an arrangement called a molecule. The smallest amount of water possible is one molecule, which is so small it can only be seen with very powerful microscopes. In fact, the tiniest visible droplet of water contains millions of molecules linked together.

Steam rises from thermally-heated water vents on Deception Island, Antarctica.

Vapor. This form of water has no shape or cohesion. It is created when water molecules are heated to such fast and furious movement that the bonds linking them break apart. This causes individual molecules to drift away from each other. If chilled, the molecules slow sufficiently to come together again. A more tangible state results: liquid water.

CHEMISTRY
It's Soup

Liquid. In the liquid form, each water molecule has stronger, more numerous bonds with its neighbors. Chilling the water tightens the bonds even further. The result is a form of water that has an even firmer shape and more cohesion: ice.

Ice. This form is cold, hard, and rigid because each of its water molecules has many firm bonds with its neighboring molecules. In fact, ice is composed of many tightly-packed molecules that have formed six-sided crystals. Heating ice causes those molecules to break apart again. Looser forms of water result: liquid water or water vapor.

What Causes Water to Change Form?

METEOROLOGY
Making Weather

In nature, water changes form over and over again, but it is essentially the same water that has been present on Earth since the planet's creation. Water can exist as water vapor, liquid water, or ice, but regardless of its temporary appearance it has been heated by the Sun innumerable times. It has evaporated into vapor innumerable times. It has condensed into rain over and over and over again. It has frozen into snow and ice almost as often. Rising temperatures have melted it back into liquid water many, many times, and the liquid water has evaporated just as often. What drives these changes?

The dominant force in water's change of state is energy from the Sun. Solar radiation sweeps the Earth's surface every day, but in a very uneven way. The Earth is a sphere. Its equator bulges out toward the Sun, so the planet absorbs most heat around its waistline. This means that waters in the tropics and lower latitudes warm up most, intensifying evaporation. In contrast, less solar energy reaches the white, ice-covered polar regions, especially during the winter months. There, water is not warmed to the point of evaporation as often. However, moving masses of air (called winds) and moving masses of water (called currents) do distribute the Sun's heat to a great extent. The worldwide movement of warm air and water (which rise) and cool air and water (which sink) serves to even out temperatures around the globe

PHYSICS
The Reasons for Seasons

How Will Water Contribute or Respond to Climate Change?

Water Vapor

Climate models predict that warmer temperatures will cause increased evaporation of water into the air. Stopping to think about wet bathing suits hung outside to dry will confirm that this is possible. Do damp suits and towels dry faster on scorchingly hot days or on chilly days? Heat speeds evaporation of water from your wet things, so they dry most quickly on a hot day (or with the aid of a good hot-air clothes dryer). With respect to global climate change, higher temperatures could contribute to a cycle of greater evaporation, increased humidity, even higher temperatures, and so forth.

METEOROLOGY
Going, Going, Gone

You might once have stepped outdoors to find the air touching your skin almost palpably damp, thick, and sticky. If so, you were feeling water vapor in the air. The hotter the air around you gets, the more water it can hold. For instance: one cubic yard (or meter) of air can absorb up to a half-cup (125 grams) of water at 15.5°C. At 40°C, the same cubic yard can hold more than two cups (500 grams) of water. Humidity is the measure of water vapor in the air. Relative humidity compares the amount of water vapor in the air to the maximum possible at a temperature. So, if the relative humidity is 50 percent, the air is holding half the water vapor it can. If it is 100 percent, the air has absorbed all it can. No more water vapor can enter the air unless (a) the temperature goes up or (b) some water vapor condenses out of the air, thereby freeing space for more.

Liquid Water

Increased water vapor in the atmosphere could cause more clouds to form. This might affect climate in two opposing ways. Some clouds take the form of low-lying layers. They spread out like sheets, moving along lazily or twisting and turning in violent motion. These clouds tend to reflect sunlight back into space, and have an overall cooling effect. In fact, it's thought that low clouds reflect back into space about 20 percent of the sunlight that reaches our planet. In contrast, high clouds, which can tower miles into the skies, were believed to warm the Earth by trapping heat that rises upward from land and waters. Recently, however, scientists have shown that thick high clouds also scatter incoming sunlight back into space, thereby partially countering their heating effects.

Does it seem confusing that clouds cool and warm simultaneously, almost as though an air conditioner and a heater are being run at the same time? Cooling or warming varies according to clouds' altitude and composition. To measure these characteristics, space-suited pilots in modified planes drop probes into clouds. Ship-launched weather balloons and Earth-orbiting satellites are also used to study clouds and their properties. As of now, these measurements suggest that clouds are reflecting away more energy overall than they are trapping. In other words, the cooling effect is dominating. But scientists can't say if this will continue to be true. If Earth continues to warm, there will almost certainly be changes in clouds' thickness and altitude, their global distribution, and their total coverage. In the future, clouds might exert a much different effect on global temperatures than now.

How else might water respond to predicted climate change? If our atmosphere warms, the volume of water in the oceans might increase and sea levels might rise. This

Anvils atop Cumulus Congestus clouds indicate the presence of ice in the clouds' upper region.

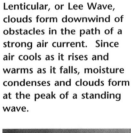

Lenticular, or Lee Wave, clouds form downwind of obstacles in the path of a strong air current. Since air cools as it rises and warms as it falls, moisture condenses and clouds form at the peak of a standing wave.

volume increase would be due to two factors. First, if the atmosphere warms, the oceans will warm as well, and water expands as it warms. Second, much water now locked into polar ice caps would melt into liquid form. Some scientists predict oceans levels might swell by 7 to 30 centimeters or more by the middle of the next century. Rising ocean levels may flood low-lying areas, and urban centers like New Orleans, Bangkok, Karachi, and Venice might suffer damage from high water levels or storm surges. If warming oceans also intensify rainstorms by speeding evaporation and condensation in the overlying atmosphere, water levels would rise even further. This would be particularly true if severe storms, such as hurricanes and typhoons, became more frequent.

Additional harm could come to coastal areas if salt water from rising sea levels invades groundwater aquifers. Many plants and most animals (including human beings) depend on the fresh water that fills underground wells and water tables, river beds, reservoirs, and lake basins. Contamination by salt water would certainly affect the potability of these water resources. If sea levels infuse coastal wetlands, marshes, and estuaries to the point of flooding, plants and animals unable to adapt or relocate would die off.

A stratospheric balloon, here being inflated prior to an early-morning launch, can carry instrument payloads weighing more than two metric tons. At 30 to 40 kilometers above Earth's surface, scientists can measure cosmic radiation, chemical reactions in the ozone, and the effect of human-manufactured pollutants.

Ice

From the window of an airplane, a field of fluffy white clouds appears almost painfully bright. That's because many clouds contain tiny ice crystals that form at freezing cold, high altitudes. It is partly due to these ice crystals that thick high clouds reflect away so much solar energy—again, approximately 20 percent of incoming sunlight. If warming temperatures change these clouds' icy composition, their cooling effects might lessen. Planetary temperatures could then warm even more.

Inside many clouds, ice crystals are continually colliding into each other or into water droplets. If enough adhere to each other, they become heavy enough to drop from the cloud. When they pass through cold air, they reach Earth in their solid state: as snow, sleet, or hail. As such, they accumulate on the huge tracts of ice and snow at our

This floating buoy measures upper-ocean temperatures and sea-level pressure from various locations on the ocean's surface to 300 meters depth. Collected data are relayed to receivers on polar-orbiting satellites.

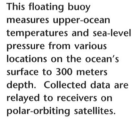

northern and southern poles that contribute to surface cooling. Climate change would affect this icy precipitation, because crystals that fall through warm air melt into raindrops.

Today, land and sea ice give parts of our planet a sparkling white surface. This surface reflects into space approximately 10 percent of the sunlight that reaches us, and helps to cool the Earth. If the world warms sufficiently, these ice packs will melt. Formerly reflective surfaces will absorb sunlight, heating Earth even more. What might happen then? If surface temperatures warm only one to two degrees, regional snow and ice might melt early, in late winter or early spring instead of in summer when water supplies are most needed. At several degrees warmer still, polar ice will begin to melt substantially. If this happens, sea levels could rise many meters over one hundred to thousands of years.

Water in all its states will almost certainly influence and be influenced by global climate change. However, the timing and magnitude of these changes are unknown. Even after decades of study, scientists are still learning about

Polar regions may experience strong effects in the event of global warming. Temperatures at the Arctic are predicted to rise four degrees Celsius for each increased degree at the equator.

the world's waters. What are all the factors that regulate the three forms of water? How might they respond to new or additional atmospheric gases, more pollutants, altered air and ocean currents, or rising temperatures? How might they respond to a changing climate? In trying to forecast our future, scientists are considering the part that water will play. In doing so, they must also consider agents capable of causing great change to the Earth's waters—and to our lands and air as well. They must consider living organisms: plants and animals, including human beings—us.

When temperatures fall, ice and snow cover more of Earth's surface. These reflect sunlight back into space.

5

Life and Climate Change

How might living organisms contribute or respond to climate change?

Plants make life possible on this planet. They are the basic building block in the food chain. They are the world's chief and most critically important food source for almost all animals, including human beings. Plants also help regulate the composition of the air around us by taking up waste gases, processing them, and releasing oxygen. In doing this, they make our air breathable. Plants shade much of the land surface, thereby cooling it. They also moisten the ground and air around them through a process called transpiration. (Plants absorb a great deal of water, but only a small amount of it is retained. Most diffuses out to the surface and evaporates, or *transpires*.) If the atmosphere changes through the addition of greenhouse gases, and if temperatures rise, what will happen to plants?

Clearing of ground cover by means of slash-and-burn techniques forces great amounts of gases such as carbon dioxide and methane into the atmosphere.

Researchers are just beginning to understand how plants react to higher levels of greenhouse gases in the atmosphere. For instance, the leaves on trees and other plants absorb CO_2 from the atmosphere during photosynthesis. Plants use the C from CO_2 to build their cell walls, so higher levels of CO_2 in the air could change the ways plants develop.

Crops versus Weeds

Plants are considered crops if they are cultivated and harvested. In contrast, plants are called weeds if they are considered a nuisance. Botanists have shown that some plants benefit from extra greenhouse gases in the air—but are those plants useful to farmers, or are they nuisances? In some cases, weeds overtake the crops with which they grow. Velvetleaf, a weed in corn fields, develops more quickly than corn when both are exposed to increased CO_2. Luckily for farmers, though, not all crop-weed combinations behave that way. In the case of soybeans, another important cash crop, researchers found that increased CO_2 in the air around soybeans causes the crop to grow taller than neighboring weeds.

In fact, many plants exposed to more CO_2 grow larger, but bigger doesn't always mean better. Though larger, they might be less nutritious, even for insects. Their leaves often contain less of the protein insects need to survive. Some insects, like the caterpillar of the soybean looper moth, might compensate by eating more plants. Others may not be able to adjust; they could mature more slowly, even dying before they're mature enough to reproduce. That might seem like good news if the insects are considered pests, but any reduction in an insect population might affect an entire food chain. If fewer

PLANT BIOLOGY
Plant Power

insects survive as a food source for birds, bats, and other animals, those animals might be put at risk. Researchers have observed this effect with buckeye butterflies, an insect eaten by some birds.

To estimate the impact of global warming on future harvests, scientists build computer simulations, or models, of climate, crops, and market conditions. First, the models are tested with present-day conditions. Then they're used to predict possible future changes. Though still in development, these models currently predict that global climate change would bring significant shifts in temperature and rainfall patterns. Though varying from region to region, such shifts would undoubtedly alter large-scale food production and international trade.

As 60 percent of the world's biomass is found in the tropics, nonindustrial pollution from this area can have global atmospheric effects.

Trees

Transpiration from trees and other plants helps cool the land surface in the same way perspiration cools our bodies. This is one way trees keep moisture in soil, especially if dropped leaves prevent dampness from evaporating. Shading the ground from the Sun is another way trees keep soil cool and moist. Today, many large stands of trees in the United States and Canada are being cleared for timber. Elsewhere, tropical rainforests are undergoing conversion into farmland and cattle ranches. These actions are controversial because native peoples, environmental groups, governments, and industry often vehemently disagree about the uses to which land should be put. However, one thing is obvious: cleared land is open to the Sun's rays. Moisture evaporates from it relatively easily. So, unless it is irrigated extensively, cleared soil dries out, erodes, and becomes less able to sustain plant and animal life.

Since living trees take up CO_2 and produce O_2, a large reduction in the number of trees might cause CO_2 to accumulate in the atmosphere. To make matters worse, cut trees burned to clear land for farming or ranching release their carbon back into the atmosphere, forming even more CO_2. Today, deforestation accounts for approximately 20 percent of the increased CO_2 in the atmosphere. A forested area the size of Central Park in New York City—860 acres—is razed every nine minutes in the tropics. At this rate, the world's tropical rainforests will no longer exist in our children's lifetime. In fifty years, they will have been destroyed.

ANIMAL BIOLOGY
On the Move

Animals

Generally, changes in climate have taken place gradually, over thousands or millions of years. Over these long time periods, natural systems had the time they needed to adjust to changing conditions. Many animal populations either adapted to new conditions, travelled in search of comfortable new homes, or both. Today, however, changes in animal habitats have been so rapid and dramatic that even large communities have been affected in some cases. If human-caused environmental changes and destruction of habitats continue at the present pace, many species could be stranded, unable to adapt quickly enough to changing conditions. Some will become extinct.

Natural and human-created climate changes can alter food available to grazing animals, resulting in migration, unusually rapid species adaptation, or death.

Changes in water temperatures may affect coral reef ecosystems, including sea fans and other corals located in shallow tropical waters.

Aquatic Communities

Wetland habitats are a case in point, as they are already at great risk. Many coastal wetlands are hemmed in by highways, housing, and seawalls. The tarmac and concrete laid over the land for such projects has already changed many areas' drainage patterns. Many wetlands have also been degraded by pollution from faulty wastewater treatment systems. Organic matter from homes and gardens is washed into waterways. Plants absorb its nutrients and grow profusely. Eventually, excessive plant growth chokes on itself and begins to decompose, producing methane and CO_2. This makes the habitat less suitable for oxygen-breathing animals such as snails, frogs, fish, and birds. Some scientists predict that rising global temperatures will worsen this problem because aquatic plants grow even better in warmer conditions.

How might global warming affect other marine animals? Coral polyps are tiny tropical-dwelling creatures whose skeletons, cemented together by coralline algae, form panoramic ramparts, reefs, shoals, and islands. Living among these formations is a dazzling diversity of shrimps, crabs, worms, mollusks, and brilliantly-colored fish. If global warming causes a rapid rise in sea level, corals supporting this animal life might be too far submerged to receive necessary sunlight. Increased rainfall, another aspect of climate change, might further harm coral and its inhabitants by washing more fertilizers, sewage, and silt into the sea. Lastly, if the ocean waters themselves warm too much, coral polyps will be unable to breathe or to feed. Warming releases the oxygen dissolved in water, and stresses the tiny algae on which corals and other animals feed.

Terrestrial Communities

Have you ever noticed that the air gets colder as you hike closer to the top of a mountain? In fact, if it is high enough, a mountain represents an entire series of temperature zones. At the base of the mountain, in the grasslands, live animals adapted to moderately warm temperatures: bison, pronghorn antelope, the western meadowlark. Farther up, in the forested areas, live bears, deer, the pine marten, woodpeckers, and the goshawk. At the highest, coldest altitudes live the animals of the alpine zone: bighorn sheep, rodents, and the regal bald eagle.

As global warming occurs, it causes an upward shift in this zonation. Temperatures and habitats now characteristic of grasslands extend up into forest lands. Typical forest conditions, in turn, push into alpine areas. What happens as the alpine zone is taken over? Does it shrink or disappear? What

Slowly rising sea levels are consistent with predictions of gradual global warming. Defenses against flooding may have to be built to protect threatened coastal areas.

happens to the animals of each zone as they are displaced by those moving up in search of cooler regions? Examples from the history of human settlements and colonies suggest some answers to these questions.

Over millennia, regional climatic variations have been severe enough to change and disrupt human communities. For instance: for two thousand years (from 4,000-2,000 B.C.), the Sahara was grassy, well-watered, and abundant with animal life. Ancient rock paintings from caves in Algeria picture animals that the area can no longer support: the water-loving bison and hippopotamus, elephants, lions...and human hunters! Over time, the area slowly became hotter and drier. Today the Sahara is a vast, sandy desert in which most water-dependent animals cannot survive.

Land degradation often involves human activities as well as naturally occurring climate change. In the case of what is now the Sahara Desert, over-farming and deforestation combined with natural causes to transform the area over the course of three thousand years. North Africa was once the breadbasket of ancient Mediterranean civilizations. Intensive farming practices caused nutrient depletion in the land. The area was also a prime source of timber, and it saw frequent warfare. In short, human-created changes in the region exacerbated a natural warming cycle.

GEOLOGY
Digging It

By contrast, Greenland warmed, then cooled. From 950-1250 A.D., Europe and the North Atlantic experienced considerable warming. Much arctic ice melted, allowing the Vikings to sail north from Scandinavia to Greenland. (They also sailed across the Atlantic to North America hundreds of years before Christopher Columbus made his own voyages.) Although it was never really very green at all, Greenland was given its name to entice settlers to migrate there. The area's climate gradually cooled, however, and Greenland became locked in sea ice. Isolated and far too cold to sustain farming, settlements gradually died out.

Aware that global climate change can have severe consequences for life on Earth, many climate specialists and engineers are studying ways in which people can adapt to changing conditions. Despite measures we now take to reduce greenhouse gases, some global warming has already occurred. Additional warming might alter important aspects of life on this planet. Scientists and engineers are working to understand and predict how climate change will progress, and how these changes might challenge human societies. They continue to search for solutions to the serious problems posed by climate change. Examples of current research worldwide include:

• *At the National Center for Atmospheric Research in Colorado, scientists are studying how rising temperatures might alter cloud cover and rates of warming.*

• *Researchers at the University of East Anglia in England are examining how various sorts of air pollution might affect global warming.*

• *Botanists at Ben Gurion University of the Negev in Israel are studying new plant strains that resist heat and drought.*

• *At Scripps Institution of Oceanography, University of California, San Diego and Max Planck Institute in Germany, scientists are investigating how global warming could change major ocean currents and climate.*

6

Human Beings and the Greenhouse Diet

What measures can be taken to limit greenhouse gases?

What changes can we make in our daily lives?

Global warming may seem like an overwhelming problem, too complex for non-scientists to understand, much less do anything about. You might wonder if one person, or one child, can help slow global warming. Can any one person make a difference? The answer is an affirmative, YES! Individual actions add up very quickly. Whenever you and your family save energy (or use it more efficiently) you save money. You also reduce worldwide use of fossil fuels such as coal and oil. This helps decrease the amount of heat-trapping gases we put into the atmosphere.

Think about this: The United States releases an average of 18,000 kilograms of CO_2 per person every year. If you multiply this quantity by the nearly 260 million people in the United States, you can calculate how much carbon dioxide the country as a whole adds to the global atmosphere. Simply lowering gas emissions by 2 percent a year nationwide would cause us to lose about 3,180 kilograms of CO_2 per person in ten years. That's significant weight loss on a Greenhouse Diet!

To see ways and amounts by which even one person can reduce greenhouse gas emissions, fill out the Greenhouse Diet chart on page 45. Though the savings given for each item have been averaged over the entire population of the United States, you'll be surprised at how much weight can be worked off at your home, at school, and at places of work.

Beyond actions each of us can take individually, there are many positive effects to be gained from becoming generally informed about environmental issues. Make careful choices beyond your home, school, and work. For instance, write and call legislators to express your concerns. Support measures that:

- *Pass laws to limit emissions of heat-trapping gasses.*

- *Set higher fuel-efficiency standards for new cars and provide incentives to buy energy-efficient cars.*

- *Provide mass transit systems to reduce travel by car, provide incentives to ride-share or to use mass transit, and encourage businesses to locate near mass transit routes.*

- *Fund research and development of renewable energy sources.*

- *Reward industries that switch from coal and oil to alternative energy sources.*

- *Direct international aid to increase the energy efficiency of developing countries.*

- *Support and fund reforestation programs worldwide.*

Of course, even beneficial actions may have side effects or tradeoffs. For example, buses emit diesel fumes, and subway systems run on electricity often generated by coal-burning power facilities. By the same token, fueling automobiles with natural gas instead of gasoline does not eliminate the emission of harmful pollutants. In some cases, alternate energy sources simply shift the burden of pollution. Solar arrays are cleaner than many existing energy sources because they have no emissions and their energy source— the Sun—is virtually inexhaustible. However, the production and eventual disposal of solar arrays usually entails the use of hazardous chemicals that must be handled carefully to avoid polluting the air, water, and ground. When evaluating alternative energy sources, therefore, keep in mind that all have costs as well as benefits.

THE GREENHOUSE DIET

What Each of Us Can Do	Home	Work	School	CO$_2$ Saved	Your Savings Per Week
Make use of recycling programs.	X	X	X	4 kg. per kg. of paper	
Walk, bike, carpool or use mass transit whenever possible.	X	X	X	3 kg. per liter gasoline saved	
Clean or replace air filters as recommended by manufacturers. Cleaning one filter can save 5 percent of the energy used to run appliances and vehicles.	X	X	X	35 kg. per air filter	
Run dishwashers only when full. Use energy-saving setting rather than heat to dry dishes.	X	X	X	90 kg. per 10 loads	
Buy energy-efficient compact fluorescent bulbs for most-used lights.	X	X	X	115 kg. per bulb	
Install low-flow shower heads to use less hot water.	X	X	X	120 kg. per 50-liter shower	
Recycle coolant whenever car, truck, or bus air conditioners are serviced.	X	X	X	225 kg. per 40,000 km.	
Drive vehicles that get good gas mileage.	X	X	X	245 kg. per 40 more kpl	
Don't overheat or overcool rooms. Lower thermostats in winter; raise them during the summer.	X	X	X	205 kg. per 1°C change	
Turn down the water heater thermostat; 49°C is hot enough for most tasks.	X	X	X	410 kg. per 10°C	
Wash laundry in warm or cold, not hot, water.	X	X	X	225 kg. per 2 loads	
Reduce waste! Buy minimally packaged goods. Reuse and recycle containers.	X	X	X	455 kg. per 25% reduction	
Caulk and weatherstrip around doors and windows to plug air leaks.	X	X	X	455 kg. per 8-10 aperatures	
Wrap water heaters in an insulated jacket.	X	X	X	455 kg. per heater	
Purchase recycled materials.	X	X	X	10 kg. per week	
Insulate walls and ceilings to save about 25 percent in heating bills.	X	X	X	910 kg. per building	
Select the most energy efficient appliances.	X	X	X	1,360 kg. per 10 appliances	
Plant trees to absorb CO$_2$ and provide shade.	X	X	X	1,360 kg. per ten trees	
Paint buildings light colors in warm climates or dark colors in cold climates.	X	X	X	1,360 kg. per building	
Install heat- and energy-saving windows.	X	X	X	4,535 kg. per 8-10 aperatures	
Request an energy audit to locate and fix poorly-insulated or energy-inefficient spots.	X	X	X	4,535 kg. per building	
Invest in energy-efficient equipment and manufacturing processes.	X	X	X	5,440 kg. per building	

Total _____

x 52 Weeks _____

Total Savings/Year _____

Ideas and Inventions

People in many countries are working to halt our movement toward a warmer world. Here are some examples of excellent ideas and clever inventions. If you can think of additional ideas and inventions, research and try to realize them. Your contribution might someday be a crucially important one to the health of our planet.

Alternative Energy Sources

Some 20 nations have pledged to stabilize CO_2 emissions at 1990 levels by the year 2000. Among them are Zimbabwe, the United States, Sri Lanka, and India. In India, for instance, the village of Pura has worked with local universities to develop a community power plant that converts manure to methane for fuel. This plant generates electricity for the entire village. Governments around the world are funding research into renewable energy sources, such as wind and solar power, that would be cleaner than existing sources.

Reforestation

Halting deforestation and planting trees on a large scale will help offset the rise in CO_2 emissions due to human activities. In Kenya, a citizens' group called the Green Belt Movement has planted ten million trees in the past 18 years. Applied Energy Services, a U.S. power company, has pledged to plant 52 million trees in Guatemala to absorb the same amount of CO_2 that its new coal-fired plant in Connecticut will release. Once planted, these trees must be protected from harm.

Population Control

By the year 2025, current world population is expected to have doubled, with the greatest increase in developing nations. This growing population will require enormous resources, including energy. To sustain this expanded global community, we must undertake one or all of the following: increase available food, housing, and energy sources; raise developed countries' average standard of living; and arrest the growth of the world population. Denmark, Germany, and Guatemala have population growth rates near zero, and other countries report negative population growth. In Thailand, improvements in the education and status of women, along with government-sponsored family planning programs, have made it possible to decrease the national birth rate by 50 percent in the last 30 years. Cuba has attained one of the lowest population growth rates in the developing world.

Water Conservation

Higher temperatures may alter the Earth's water cycle. Some regions would get more rain, while others, particularly the mid-regions of continents, might become much drier. New irrigation systems may have to be built for drought-stricken areas. Increased conservation efforts are being planned to stretch existing supplies along the southwest coast of Australia, the west coast of China, and in South Africa, Swaziland, Brazil, Uruguay, and Mexico.

Protection from Rising Seas

If global warming causes sea levels to rise, dikes, sea walls and other defenses against flooding will have to be built to protect threatened coastal areas. Work along these lines has already begun in the Unites States, Guyana, Brazil, Gambia, Spain, the Netherlands, Israel, Australia, and Thailand.

Passageways for Wildlife

Many animals may adjust to climate change by moving to new habitats. Their travel routes could be blocked, however, by farms, roads, homes, and cities now surrounding many forests and wildlife reserves. Travel corridors could be provided for animals by building tunnels under highways. Setting aside strips of undeveloped land could also help animals to bypass cities and other obstacles. Wildlife reserves could be enlarged to give some species more room to move within protected areas. In the absence of such measures, many species could become stranded in unsuitable habitats and die out. Canada, Cuba, Haiti, Argentina, Rwanda, France, Malaysia, and Japan are leading efforts to develop pathways and refuges for animals.

Agricultural Changes

Climate change might benefit farming efforts in some areas, but in others, agriculture might suffer. New agricultural practices will be needed to cope with changes in temperature and soil moisture. Farmers might begin growing more heat-tolerant strains of their present crops, or switch to different crops. They might also need to develop measures to control invading warm-climate insect species and weeds. New Zealand, Japan, Vietnam, Zambia, Ethiopia, Nigeria, Scandinavia, Canada, the Great Lakes region of the United States, Guatemala, and Chile are already undertaking these efforts.

On the Move

ANIMAL BIOLOGY

Background information

Locate several thermostats around your school and home. At what temperature are they set? Most people feel best when the air temperature is between 18 and 24°C. When temperatures drop much below 18°C we put on warm clothing or turn up the heat. When temperatures are too high, we jump into the ocean and swimming pools or turn on fans and air conditioning. In short, human beings adjust to the natural range of temperatures that occur from day to night and from season to season. Other animals make similar adjustments. Furry animals grow thicker coats to keep warm in the winter. Many animals seek the comparative coolness or warmth of caves, cracks, or crevices. Birds and fish leave home altogether in search of warmer or cooler zones.

What might happen if current temperatures change dramatically from what we think of as normal conditions? What will happen to penguins if it warms up so much that the ice on which they live melts? What might happen to a tropical parrot if rainforest temperatures dip below freezing? Would these animals be able to tolerate the stress? Would they move away? Would they die? Such extreme changes are not likely to take place soon. If global greenhouse warming occurs, however, local communities will almost certainly experience shifts in climate. Changes in temperature, humidity, or rainfall might restrict the availability of areas in which a particular animal can survive. Temperature changes affect most animals indirectly as well. For example, the abundance of an animal's food source could be altered by temperature changes. Changing conditions might also favor or hamper predators whose role it is to keep animal populations from increasing unbounded.

In response to changes in their environment, some animals will be able to migrate to new areas, but other animals will be prevented from moving by barriers. (Examples of barriers include rivers, oceans, mountain ranges, and various structures such as roads, fences, and housing developments.) To count the rising or falling numbers of animals living in a given area over time, scientists use a variety of techniques. It is generally not possible to survey a very large area, so scientists *sample* the area—they study one portion of a region considered to be representative of the entire area. This often involves the use of quadrats, or squares that enclose a portion of the survey area. The number or type of animals in each randomly-placed quadrat is considered to be representative of the whole area. For example: if an area to be sampled is 10 square meters, and a quadrat is one square meter, then the number of animals living in the whole area should be about 10 times the average number found in the quadrats. The difference between this estimate and the true number is known as the standard error. (The standard error is scientists' way of acknowledging that measurements might be off by some small amount. Nobody's perfect!)

Objective:
To investigate the effects of climate change on animal distributions.

Materials:

20 cm X 20 cm card stock or stiff paper

scissors

large box or aquarium

sawdust sufficient to fill box or tank to depth of 13 cm

cardboard divider

masking tape

200 mealworms (from scientific supply house or pet store)

blindfold

large spoon or ladle

4 paper plates

pen or pencil

heat lamp

Introduction to activity

a. List things that animals and plants need to survive. Discuss how animals might be helped or hurt by a rise in temperature. List the things that animals do to cope with heat.

b. Brainstorm how specific animals have evolved to survive under certain circumstances.

c. Discuss why scientists cannot count and categorize all animals in an entire area of interest, such as North America or the Pacific Ocean.

d. Demonstrate random sampling techniques. How are sampling areas chosen? Choose a large area such as the front of the classroom and divide it into 20 or more numbered squares. How might you randomly select four squares?

Procedure

a. Cut 4 squares, 5 cm on a side, from card stock. Each quadrat's surface area is 25 cm².

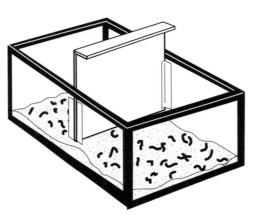

b. Spread sawdust over the bottom of the aquarium or box. Divide the box into two equal halves with the divider. Tape the divider to the sides, leaving a small gap about 2 cm high at the base for the animals to pass through.

c. Scatter the mealworms evenly throughout the box. Count how many are added to each side.

d. Blindfold 1 student. Have him/her gently place 2 quadrats on each side of the box in random positions on top of the sawdust. Remove the blindfold and record each quadrat's location on a separate paper plate. Use a pencil to make a line in the sawdust around the quadrat cards.

e. Carefully lift 1 quadrat without disturbing the lines in the sawdust. Using the ladle, remove the underlying mealworms and sawdust to the plate labeled with the quadrat's location. Repeat with the other three cards.

f. Examine each sample for animals. Count the number of mealworms in each, and record the number on the data sheet. After counting all the samples, return the mealworms and sawdust to the same half of the box from which they were collected.

g. Position the lamp above one half of the box, according to your teacher's directions, and turn it on. Gather predictions about changes that might occur as one half of the box warms.

h. After the box has been warmed, repeat the sampling procedure (steps d through f) in each half of the box. Calculate the average number in each side of the box, based on the number in your samples. Note the changes in abundance.

Discussion

a. What happened to the distribution of animals in the box?

b. Was the quadrat sampling a good measure of the true number of animals? Were the estimated numbers close to the number originally placed in each side of the box? How could sampling be improved?

c. What types of animals would do better in a warmer climate?

d. How do plants differ from animals in their response to warming?

e. This demonstration allowed animals to move when conditions did not suit them. What would happen to animals that cannot move away?

Example:

If 7 mealworms were found in one quadrat and 5 in the other quadrat in the same half of the box, then the average is *(7 + 5) ÷ 2 = 6.* The estimated number in each half of the tank is the average number in the area of a quadrat multiplied by the number of quadrats that could fit in the whole half. That is, if each half of the box measures 50 cm by 20 cm, the area is 50 x 20 = 1000 cm^2. The quadrats are 25 cm^2, so 20 quadrats could fit in each half of the tank *(1000 ÷ 25 = 20)*. The total estimated number of animals in each half will be the average number per quadrat (6) multiplied by the number of quadrats per half (20), so we estimate that there were 120 mealworms in each half of the tank.

Teacher's Guide to the Activity

Time:

This activity takes 15 minutes to prepare, and approximately 30 minutes to complete, with a break of several hours or a day.

Because scientists cannot accurately survey a very large area, they select a representative portion, or sample, to analyze. In this sampling activity students are asked to observe the response of mealworms to a change in temperature. Using a box or aquarium divided in half, students observe the response of mealworms before and after a rise in the temperature. A blindfolded student selects at random four sections, or quadrats, to measure the distribution of mealworms.

Hints

• An incandescent lamp with a flexible arm or a clamp that attaches to a ring stand is suitable for this activity.

• Pretest the lamp and box setup to determine how fast and by how much the temperature will rise. A "cooler" heat source may allow the experiment to progress overnight. A warmer heat source may heat both sides of the box if left on too long.

Introduction Guide

a. To cope with heat, animals might sleep during the daylight hours, drink lots of water, pant, sweat, or bathe.

b. Ducks have webbed feet to facilitate swimming. Mountain lions' great speed enables them to catch prey. Birds and butterflies have coloring that helps them blend in with their surroundings. Seals have a layer of blubber for insulation. Hawks have keen eyesight so they can spot their dinner from high altitudes.

c. Such extensive studies would take too long, cost a great deal, upset animals, disturb the environment, or simply be impossible.

d. Scientists try not to let biases or judgment dictate their selections. Instead, random processes determine placement of sampling devices. Demonstrating random sampling techniques can be done in two ways:

1. Blindfold one student, and have another student lead him/her around the area. When the blindfolded student says "stop," the square occupied by his/her left foot will be "sampled." Select four squares to be sampled.

2. Use a random number generator such as dice or the numbers 1-20 on pieces of paper placed in a hat to choose four squares that will be sampled.

Discussion Guide

a. Many or all of the animals should have moved to the cool half of the box. Worms in the heated half of the box may have died.

b. Differences in measurement might be due to procedural (standard) error, such as the death of some worms due to handling. To increase measurements' accuracy, the number of quadrats could be increased.

c. Animals that cannot regulate their body temperature, such as reptiles and insects, would survive better in a milder climate.

d. Plants cannot move. If a plant species is to survive, its seeds must reach suitable places for germination.

e. Animals that cannot move away from new conditions will either accommodate or die. Adaptation to changed conditions often requires many generations. Climate change might take place quickly enough that animals can't keep up. For example, if a coastal habitat is flooded by rising seas, land animals who live there can't rapidly develop gills so as to breathe underwater. These animals might be able to move away from the rising waters onto higher ground, but they might encounter survival problems in their new location. They might not find adequate food, or they might be out-competed for food and territory by other animals who already live there.

ANIMAL BIOLOGY

A Fishy Tale

Objective:
To learn how fish scales can be used as indicators of changing environmental conditions.

Background information

Have you ever touched a fish and noticed that its surface is smooth in one direction and rough in the other? Most fishes are covered with scales, each overlapping the one behind it, that protect and streamline the fish. A scale has two layers: a thin, inner layer of tissue connected to the skin, and an outer bony layer. The bony layer is made up of concentric ridges showing growth increments during the life of the fish. Spacing of these ridges, called "circuli," gives biologists clues to the life history of the fish.

As with tree rings, which add a layer each growing season, fish scales have annual rings. Year marks, called "annuli," can be detected on many scales by skilled scale-readers. Annuli occur where circuli are clustered together. The spacing between annuli gives an indication of the particular conditions the fish encountered during a period of growth. For this reason, it was once thought that growth records as shown in scales would reveal how environmental conditions have changed over time. But it is now clear that fish scale rings are unsatisfactory indicators of climate change. First, so many factors influence ring width (*e.g.*, food availability, fish migration, varying growth responses to temperature) that it is nearly impossible to attribute changes solely to climate change. In addition, individual scales represent relatively brief amounts of time; they generally show years rather than decades or centuries of growth. Therefore, any given scale shows scientists only a partial picture of changing environmental conditions. Collections of many scales, such as those that settle to the seafloor over decades or centuries, are far more useful.

A fish keeps most of its scales for its entire life, but some scales are lost and replaced. Discarded scales drift downward and accumulate in sediments. These collections of scales serve

A simplified fish scale. Scales from bony fish you will examine are likely to be a little different from this diagram. Common bony fish such as perch have scales that tend to be flatter in the anterior part (closest to the head). Posterior (back) parts of the scale have tiny spines.

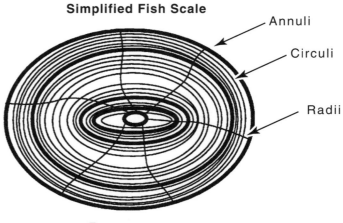

Simplified Fish Scale

Annuli

Circuli

Radii

Focus at center

as records of environmental change. By identifying the origin and number of scales in the layers of a sediment core, scientists can infer which fishes lived in a specific area at particular times. Changes in fish populations can be very revealing because different species of fish prefer particular ranges of water temperature in the same way many land animals prefer cooler or warmer habitats. Shifts in scale deposits suggest changing conditions. Off the Southern California coast, for instance, sediments deposited in warmer periods have greater abundances of sardine scales, while sediments deposited in colder times tend to have greater abundances of anchovy scales.

Recently, some scientists have used fishes' otoliths (ear bones) rather than scales to search for indications of changing environmental conditions. Fish develop daily as well as annual layers in their otoliths, so layers in fossil otoliths can be used to obtain a detailed development record. (It is possible to see otolith layers under a dissecting microscope, particularly in still partially-transparent otoliths from young fish.)

Measuring growth increments in fish scales

Introduction to activity

a. Fish in warming waters grow faster or more slowly, depending on the predisposition of their individual species. Changes in salinity (the amount of salt in the water) also influence growth rates. What other climate changes might affect fish growth?

b. Many living things respond to climatic conditions. Some crops, such as mangos, pineapples, and bananas, grow best in warmer conditions. Other crops, including cabbage, potatos, and squash, thrive in cooler conditions. Check newspaper ads, visit local supermarkets, or interview a greengrocer to obtain the prices of selected produce over the course of a year. How and why does the abundance of fruits and vegetables fluctuate in response to seasonal and other conditions?

Procedure

a. Remove several scales from the fish with the tweezers. Wrap scales in a moist paper towel when they are not being examined.

b. Place a scale under the dissecting microscope and look for the following features: circuli, annuli, radii, focus.

c. Draw and label an illustration of the scale that you see under the dissecting microscope.

d. Look for a scale that has very large growth increments. This might be a replacement scale—a scale which grew very quickly to replace a lost scale.

Materials:

at least one large fish, or a section of a fish, with scales intact, *or* scales collected from fishmonger

1 pair of tweezers

paper towel

water

clear plastic or glass microscope slide to mount scales

1 dissecting microscope (x20 magnification)

1 ruler

diagram of an idealized fish scale

paper

pencils

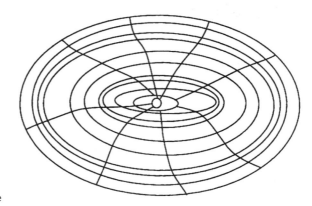

**Diagram of an
idealized fish scale**

e. On the diagram above, measure the distance between the annuli. The circuli are not shown, to make the diagram clearer. Measure distances by drawing a line from the focus to the outside of the scale and measuring the width of each increment (*i.e.*, each gap between annuli). Given the number of annuli, calculate the approximate age of the fish. Speculate as to the conditions the fish might have encountered in each year of its life.

The shape of this scale suggests that it comes from a fish that grows best when the water is warmer. Assume the fish was caught in 1995, when the average water temperature in this fish's habitat was 11°C. Also assume that one degree of temperature leads to 1 mm of growth difference. What temperature conditions might the fish have encountered in earlier years? The outer ring is 6 mm wide, so the average temperature was 11°C. The next ring is 3 mm wide, so the average temperature was 8°C, and so forth.

Discussion

a. We have assumed that our scales came from a fish that spent all of its life within one geographic area. How does the interpretation of measurements change if the fish migrated from one area to another? Do your measurements indicate that the fish migrated to areas of different temperature, or that temperatures in one area rose and fell?

b. If a fish grows throughout its life, each increment can be compared to any other. How would the ring widths appear if growth slowed as the fish aged?

c. Scales' orientation and the irregularity of their rings affect measurement. Calculating ring width, for instance, depends on scales' position and the direction of measurements made. What standard means of measurement can students devise so that comparisons among scales are uniform? Devise a standard way in which to measure the scales for comparison between scales.

Teacher's Guide to the Activity

All living things reflect the conditions under which they lived. An increase or decrease in growth is reflected in the scales of a fish, much as in tree rings. In this activity students study fish scales to identify changes in the environmental conditions experienced by a fish.

Hint

• In selecting fish to observe, choose several different species. Red snapper, tuna, mahi-mahi, halibut, and salmon are good choices because their scales are large enough to handle easily. Scales from other fishes also permit comparisons to be made.

Introduction Guide

a. A change in the abundance of food affects fish growth. Changes in currents and the strength or direction of the wind move fish toward or away from food-rich areas. If carried by winds from land to the oceans, nutrient-laden dust might encourage plant growth: more fish food!

b. Some crops grow and are harvested in warm conditions, while others such as apples must undergo a period of low temperatures if they are to thrive.

Discussion Guide

a. This is a difficult question to answer because there is no way to tell whether a fish experienced an overall change in climate. We can only tell that some condition or conditions changed. This change might have been due to either altered location or altered conditions in the same location. In studying climate change, we are most interested in measuring how conditions changed in the one location. Scales from fishes that live in the same area their whole life would be most useful. In fact, the wide range of fluctuating temperatures suggested by this fish's scale suggests that it did migrate. In contrast, gradual and unidirectional change suggests true climate change in a single location.

b. Ring increments do decrease as fishes age, even if the fishes' conditions stay stable. Given this, growth must be calculated against the age of the fish. This is one of the complications that scientists must consider. Fish also grow more slowly when they breed, because energy is diverted by reproductive processes. Thus a fish that reproduced could be taken for a fish that grew slowly due to environmental conditions.

c. Students should orient scales so that the head-to-tail plane is horizontal, with the head end on the left. Rings should be measured from the focus directly to the bottom of the scale.

Time:
This activity takes 15 minutes to prepare and 45–50 minutes to conduct.

CHEMISTRY

Objective:

To show that air contains invisible gases such as oxygen, carbon dioxide, and water vapor.

Air Cares

Background information

What's inside an inflated balloon? Helium? Air? Someone's breath? These substances are all invisible gases or, in the case of air and breath, mixtures of gases. Helium gas is lighter than air, so a helium-filled balloon floats up into the air when released. The Goodyear Blimp is also filled with helium, but ballast (weight) is added to the blimp so that it hovers in the air.

Air contains many gases. It is mostly nitrogen (78 percent) and oxygen (21 percent), but other, less abundant gases in air are carbon dioxide (CO_2), methane (CH_4), and water vapor (H_2O) as well as many other trace (rare) gases. Oxygen is an essential gas for most living organisms on Earth, including human beings. With every breath we inhale, our bodies extract some of the oxygen present in the air. The breath we exhale is slightly richer in CO_2 than the air we take in. CO_2 accounts for approximately 0.03 percent (one thirtieth of one percent, or three parts in ten thousand) of air. This might seem like a very small percentage, but the amount of carbon dioxide in the atmosphere has been increasing since people began burning coal and gasoline for energy. Many scientists believe that this increase in CO_2 could cause the temperature of the Earth to rise.

Like carbon dioxide, water vapor is a heat-trapping greenhouse gas. Unlike CO_2, the concentration of water vapor in the atmosphere varies greatly from place to place, and depends on temperature. Warmer air can hold more water vapor than colder air. In this experiment we will demonstrate that O_2, CO_2, and water vapor—three invisible gases—are present in air.

Introduction to activity

a. Demonstrate how a trace gas can be introduced into the air. A dish filled with an aromatic substance will be placed in the center of the room. The smell will travel through the room as its complex mixture of molecules evaporates on contact with O_2 in the air. Raise your hand when you are able to smell the substance.

b. Capture your own complex aroma by finding an aromatic substance at home and placing it in a plastic bag. Put the substance in a paper bag to hide it and challenge classmates to identify the odor-causing substance. Examples include perfume, eucalyptus leaves or pine needles, and mushrooms.

c. Find a Periodic Table of the Elements in a chemistry book or encyclopedia. Find out which gases are elements at room temperature. Choose a gas and find out how much of it is found in the Earth's atmosphere.

Procedure

Part 1: Testing for oxygen.

a. Light the short candles.

b. Place the glass jar over one of the two candles.

c. Let the candle continue to burn inside the glass jar until it goes out.

d. Slide an index card under the jar, sealing it as you pick it up and turn it over. The mouth of the jar now faces upwards. *This step must be performed quickly because the gases in the jar are now heavier than air. They will escape from the jar if its mouth is pointed down.*

e. Remove the index card and immediately "pour" the air in the jar over the second burning candle. It should go out.

Part 2: Testing for water vapor.

a. Place the mirror in the bucket of ice to cool for 5-10 minutes.

b. Remove the mirror from the ice.

c. Wipe any water off the surface of the mirror with a paper towel. Be careful not to handle the mirror so much that it warms.

d. Observe the surface of the mirror. What happens to it as it warms to room temperature?

Part 3: Testing for carbon dioxide.

a. Pour distilled water into a glass, and test its pH with litmus paper. Notice the color of the litmus paper.

b. Connect the air stone to the tubing. Place the air stone in the glass of water.

c. Have the students take turns blowing through the air stone for a total of 10 minutes.

d. Repeat the pH test with a new piece of litmus paper. Notice the color of the litmus paper. Compare it to the paper from the first test. Is the color different?

Materials:
two short candles

medium-sized pickle or other glass jar

matches

one 10 x 15 cm index card

bucket of ice

paper towels

small mirror (perhaps 10 x 10 cm)

drinking straw *or* a foot-long piece of aquarium tubing

aquarium "air stone" (from a pet or tropical fish store)

glass of distilled water

litmus paper capable of testing pH in the range of 6.0 to 8.0 (a larger range is fine)

Discussion

a. Why did the candle go out when burned inside the jar?

b. Why did the second candle go out when the air in the jar was poured over it?

c. Why did water condense on the surface of the cold mirror?

d. Why did the color of the litmus paper change after air was blown through the water?

Teacher's Guide to the Activity

This activity helps students to view air as a mixture of invisible gases by testing for oxygen (required to keep a candle burning), water vapor (condensing on a cool mirror), and carbon dioxide (forming slightly acid ionic compounds when students blow into otherwise neutral water).

Hint

• A handbook of chemistry and physics will provide further information about elemental and compound gases.

Introduction Guide

a. As an introduction to this activity, demonstrate how a trace gas can be introduced into the air. Place a dish in the center of the room. Fill it with an aromatic substance such as white vinegar. Ask students to raise their hands when they are able to smell the vinegar. Air currents are responsible for diluting and spreading the odors.

b. Suggest students bring in aromatic substances that will not rot or lose their smell. Prepared foods such as fried bacon may also be used.

c. Notice that CO_2, CH_4, and other familiar gases are not found on the periodic table. This is because the periodic table includes only elements. It does not list compounds, or molecules made up of more than one element.

Discussion Guide

a. Candles require O_2 to burn. O_2 is one of the gases present in air. Therefore, a candle will burn as long as O_2 is available. When the jar was placed over the candle, O_2 was present in the air inside the jar. However, as the candle burned, it consumed the O_2 inside the jar and produced CO_2 and other gases. Eventually the candle exhausted all the O_2 in the jar, and the flame went out.

b. The air left in the jar after the candle burned out had no O_2 in it. It was also heavier than the air in the room because CO_2, a dense gas, had been added to it by the burning candle. Thus, when this "heavy," O_2-poor air was poured over the candle, it displaced the O_2-rich air in the room and extinguished the flame.

c. Water vapor is the gaseous form of liquid water. It is formed as water evaporates from the oceans, lakes, and rivers. Warm air holds more water vapor than cold air. When the mirror's cold surface chilled the air close to it, water vapor present in the air condensed (became a liquid) onto the surface of the mirror. If there had not been any water vapor in the air, no water

Time:
Part 1 of this activity takes 10 minutes to prepare and 10 minutes to conduct; Part 2 takes 5 minutes to prepare and 20 minutes to conduct; and Part 3 takes 10 minutes to prepare and 30 minutes to conduct.

droplets would have formed on the mirror. Clouds are formed in a similar way. Warm air holds a lot of invisible, gaseous water vapor. As this warm air rises into the atmosphere, it begins to cool. This causes some of the water vapor in the air to condense into liquid water droplets. These droplets are visible, and form clouds. If these droplets become large enough, they fall back to Earth as rain.

d. Water is neither a base nor an acid. It is neutral. Thus, its pH should be relatively close to 7.0 on a scale of 1 (acidic) to 14 (basic). As air is blown through the water, CO_2 from the air reacts with the water to form carbonate ions. These are ionic compounds which increase the acidity of the water. The litmus test should reveal a slight change toward a more acidic pH, which causes the visible difference in the color of the litmus paper. If no CO_2 were present in the air, no carbonate ion would form, and no change would be detected. (Note: There is always some CO_2 in air. If this experiment does not produce the expected results, it is probably due to insensitivity in the testing procedure, not to a lack of CO_2 in the air.)

It's Soup

CHEMISTRY

Background information

"Water" is the word identifying a single molecule composed of three atoms. Water is also used to refer to a collection of molecules in bulk: what we know as water vapor (gaseous water), water (liquid water), and ice (solid water). The single molecule of water contains only hydrogen and oxygen. Collected molecules of water, in contrast, incorporate many additional substances as they come together: minerals (*e.g.*, calcium or magnesium salts, depending on whether the water comes from a lake, river, reservoir, or well); elements from principal treatment processes such as fluoride and chlorine; and gases drawn in from the air, including nitrogen, oxygen, and carbon dioxide. Technically speaking, therefore, H_2O is in essence a very weak soup consisting of a water base lightly seasoned with a variety of organic and inorganic "spices." These spices are generally invisible to the naked eye, but can be exposed.

Like any standing body of water, the Earth's oceans contain many constituents over and above water. In addition to minerals (leached from rocks and soil) and organic matter from living or once-living organisms, the oceans absorb oxygen, nitrogen, and an important green-house gas: carbon dioxide. Oceans take in CO_2 and other gases to maintain an equilibrium, or a balance of concentrations, between gases in the sea and air. How do the oceans absorb CO_2? Once drawn into water, CO_2 molecules react with already-present carbonate ions to form bicarbonate ions ($2HCO_3^{-1}$). This transformation removes CO_2 proper from the water. Further-more, bicarbonate ions are used by marine animals and plants in forming the calcium carbonate ($CaCO_3$) with which their shells are made. When these marine organisms die, their shells settle to the sea floor and are often buried in sediments there. Some fraction of the calcium carbonate dissolves back into the ocean waters, but a large portion is trapped as solid sediment, effectively locking away the atmospheric CO_2 used to make it. Through these processes, the oceans ultimately determine the atmosphere's CO_2 content.

At one time, it was thought that oceans could continue to absorb carbon dioxide indefinitely. Although the extent to which Earth's oceans can take up greenhouse gases is still not fully known, some scientists now believe that oceans absorb about half of the CO_2 produced by human activities. This might change if ocean temperatures rise due to global climate change, however, because gases in warm water become less soluble. If more readily released into the air, additional CO_2 would intensify an already-problematic greenhouse effect.

Introduction to activity

a. Discuss the composition of seemingly clear standing bodies of water, including oceans, rivers, lakes, reservoirs, puddles, and the contents of rain barrels. Create an ingredient list for these natural soups. List possible sources of each ingredient including the water itself, thereby consider-ing the hydrologic cycle.

Objective:
To demonstrate that what appears to be clear water is actually a complex solution of dissolved gases, mixed liquids, and solids.

Materials:

unopened bottle of club soda or carbonated mineral water (clear, plastic, screw-top bottle)

2 small glasses of equal size

water from tap

distilled water

boiled water (cooled)

aerated water

b. Demonstrate that water contains invisible gases. Collect and cover samples of several different types of water: tap, boiled, distilled, and aerated (shaken vigorously to force up the oxygen level). Study these by eye and using a hand lens. Are there any visible differences? Divide each sample into two equally sized small glasses. Cover one, leave the other uncovered, and place both in a sunny and/or warm spot for 24 hours. Are any differences visible now? What happens when these various waters are frozen? Which water would you use to create the clearest possible ice cube?

c. Demonstrate water's limited capacity to dissolve another substance. Add spoonfuls of salt to a glass of water, one at a time. Stir after each addition, keeping count of the number of spoonfuls added. Stop when the salt will no longer dissolve. The solution is now saturated with salt; it will hold no more. Saturation limits for various substances change with temperature and pressure.

Procedure

a. Open the bottle and listen for the characteristic "whoosh" sound of gas escaping from the bottle. Also observe bubbles forming in solution and rising upward.

b. Re-cap the bottle, then immediately remove the cap for a second time. Was a whoosh heard this time? Again, re-cap it and set it aside for several hours or until the next day.

c. Open the bottle and listen for the whoosh. Re-cap and set aside. Repeat this process until you no longer hear the whoosh upon opening the bottle after a long rest. Count the number of times you repeat the process.

Discussion

a. What causes the whoosh when the bottle is first opened?

b. Why doesn't the bottle whoosh when it is re-capped and then immediately re-opened (in step "b" above)?

c. Why does the bottle stop burping after many repetitions?

d. How does this demonstration involving CO_2 dissolved in water relate to the real world and the subject of global climate change?

Teacher's Guide to the Activity

In this activity, students open and re-seal a bottle of carbonated soda to observe audible changes in pressure and visible changes in the liquid. Although water appears "pure" to students, this activity demonstrates that what looks like clear water is actually a solution of dissolved gases, mixed liquids, and solids.

Introduction Guide

a. Ingredients may include water, carbon dioxide, oxygen, rock and bone minerals, sand particles, fish mucus and bird excrement, industrial pollutants, feathers and fur.

b. Bubbles clinging to the glass are gas pockets that were dissolved in the water and are now out of solution. When water heats up, the gases in it become less soluble and are easily released. By the time you see them, many gas molecules are no longer mixed with the water molecules. Rather, they have come together in a bubble. Freezing rigidifies water molecules, starting at the surface where there is the most cold air: the outside surface. Freezing gradually moves inward, pushing irregularly shaped constituents—*e.g.*, salts, minerals, and extra gases—into the center. Distilled water is produced by boiling water and collecting the condensed steam. Distilled water is fairly free of impurities, and creates the clearest ice cubes.

c. Students' water incorporated only so much salt. Likewise, the oceans absorb limited amounts of gases from the atmosphere, including greenhouse gases. In nature, water is not saturated with CO_2—it has the capacity to hold more. However, there is a balance of concentrations between CO_2 in the oceans and in the atmosphere; that balance would need to change before oceans could absorb more gas.

Discussion Guide

a. The solution is supersaturated with carbon dioxide. This means that at the conditions of room temperature and pressure, there is more carbon dioxide (CO_2) dissolved in the solution than it can accommodate. The pressure of CO_2 inside the bottle is greater than the pressure of CO_2 in the air outside the bottle. When the bottle is opened, some of the CO_2 gas rushes out to equalize the pressure.

b. Notice that the bottle is not completely filled with liquid; there is a small pocket of air at the top. In the unopened bottle, some of the carbon dioxide has come out of solution and entered the air pocket. The pressure of CO_2 in solution and the pressure of CO_2 in the air above the solution are equivalent. But this equalization process is not instantaneous. When you open, cap, and reopen the bottle rapidly, there is not enough time for the air pocket in the bottle to re-pressurize with CO_2.

c. Each time the bottle is opened and re-capped, a new state of equilibrium is reached. That is, the pressure of CO_2 in the liquid and in the air pocket equalize. And each time you open the bottle, the pressure of CO_2 in the air pocket and in the air outside the bottle equalize. With each repetition, some of the excess gas in the soda escapes from the bottle. Eventually, the pressure of CO_2 in all three zones—liquid, air pocket, air outside bottle—is the same, so there is no imbalance to cause the burp.

d. The burning of fossil fuels (coal, oil, gas) is causing levels of CO_2 and other heat-trapping gases to build up in the atmosphere. This increase might cause global temperatures to rise, changing conditions in various areas in ways that might be inconvenient or harmful. Some of the excess CO_2 that we are releasing is absorbed by the oceans and continents. So it is important to understand how much CO_2 can be held. Just as there was a limit to the quantity of CO_2 dissolved in the soda, there is a limit to the amount of gas that can be stored in reservoirs through natural processes.

(Gonna Take A)
Sedimental Journey

GEOLOGY

Background information

If you were going to construct a multilayered cake, you'd start with the bottom layer, place additional layers of frosting and cake on top, and frost the sides. Now imagine that you are presented with such a cake. It is so completely covered with frosting on the sides and top that you can't see the cake underneath. What would you do to find out what kind of cake was inside? You'd probably cut a wedge from the cake and examine all the layers.

Layers of rock and sediment can resemble a layer cake in structure. Sediments are particles of sand, silt, and mud that settle downward on land or onto the floors of lakes and oceans. Over time, older sediments are covered by successively younger and younger material. In each layer, evidence of past environmental conditions is locked away in what is called the geologic record. Scientists analyze the rock or sediment type, the chemical composition of the minerals, the organic material in the layers, and sometimes the fossilized remains of once-living plants and animals, particularly pollen from plants and shells and bones of animals. These data help geologists understand how climate has changed over time.

To uncover these clues geologists, like cake-cutters, must expose the layers of sediments. They do this by driving a long, hollow tube into the sediments and extracting a core. A core is a long, plug-like cylinder of sediment extracted from the bottom of a lake or ocean. The oldest sediments are those deepest in the core. The reason the sediments are layered is that conditions change above the spot where the core is taken. Scientists reconstruct climate changes by interpreting the layers in sediment cores. For example, sediment color provides clues about past ocean chemistry. Some sediment particles are rich in iron (Fe). When the seawater has plenty of oxygen gas (O_2) dissolved in it, the iron is converted to iron oxide (FeO_2) and the sediments take on a rich red-brown color. Sulfur (S) colors sediments yellow-brown. Volcanic ash might also be found in the sediments, and can be used to construct a history of past volcanic eruptions.

Perhaps most interesting of all are clues to past climate contained in plant and animal remains. A great deal of plant material in lake sediments suggests that growing conditions were favorable at the time the material was deposited. This means that sunshine and rain were probably plentiful. Colder, drier conditions are generally less favorable for plant growth, so less plant material is present from cold periods. Because unfavorable weather causes loss of vegetation, however, cold-period erosion on the shores might ultimately result in more plants being uprooted and swept or carried into waters. Given this, scientists search for more stable indications of conditions influencing plant growth. One such indication is the presence or absence of pollen grains. Because pollen has a thick protective coat that helps prevent decay, it is very well preserved in sediments. A large amount of pollen in sediment cores suggests that climate was favorable to plant growth. Little or no pollen suggests the reverse (or that unusual conditions prevented pollen from blowing into water).

Objective:
To simulate and interpret pollen-bearing sediment cores.

Materials:

1 bucket or container, at least 20 cm deep and 10-15 cm wide

1 fist-sized ball of clay

3 small buckets of sand, soil, and small pebbles, respectively

3 small bags of cornmeal, unbleached flour, bleached flour, and/ or variously colored gelatin powders

2 clear plastic syringes with the needle-holding end cut off

pencil or marker

2 rulers

Introduction to activity

a. Not all plants produce pollen. List plants that produce pollen, particularly those that are found locally.

b. Consider some ways pollen is taken to and from plants. Huge numbers of pollen grains are produced by some trees. Many pine trees have about fifty clusters of cones, together producing about one million grains. Each tree might have one hundred of these clusters. (The thick yellow dust found on cars parked beneath pine trees is pollen.) Pollen that is transported by the wind will often end up in lakes.

c. List types of plants and trees that grow in cold climates and in warm climates. Consider pine trees and other evergreens, oaks and other deciduous trees (trees that lose their leaves in winter), and grasslands.

Procedure

Part I

a. To create simulated sediment beds using a bucket, various sediments, and pollen-like powders, begin by desigining a key that shows what powders, or "pollens," represent which plants. For example, cornmeal might represent grass pollen.

b. Pat 1 cm of clay across the bottom of the container.

c. Sprinkle a layer of sand over the clay, then add pollen grains to represent one climate area. Distribute differently colored powders across the sediment layer to represent pollens from different plants.

d. Continue adding layers of sand until the container is full, alternating the colors and depth of sand, and the type and amount of pollen grains. Assume that one centimeter of sand accumulates each 100 years.

e Exchange sediment samples with another group.

Part 2

a. Extract a core from the sediment sample with the modified syringe. Insert the core vertically through the sediment layers, then carefully remove it by pulling upwards. The bottommost clay layer should plug the bottom end, preventing the sediments above the clay from falling out. Look through the sides of the clear corer. Mark the top of the core on the corer. These represent the youngest sediments.

b. Measure the depth of each sediment layer. Record this on a data sheet. How many years does each layer cover?

c. Extract each layer one at a time, and sort for pollen grains. Record the number of grains and type of pollen.

Part 3

a. Describe the changes in climate indicated by your sediment core, and check these against the key of the group that originated the sediment bed.

b. Draw or write a description of the trees that produced the various pollens in the core sample, and the possible climate for each layer in the sediment.

c. If the top layer was deposited only yesterday, how far back in time does your core extend? In other words, when were the oldest sediments deposited?

Discussion

a. What kinds of pollen would be found in lake cores taken near your house and school? What about cores extracted from lakes in different parts of the country?

b. If global warming occurs, what kinds of pollens might be deposited in lakes around the world?

c. What other materials might be deposited in sediments that would serve as clues in analyzing past climate?

Teacher's Guide to the Activity

Time:

This activity takes 10 minutes to prepare and 50–60 minutes to conduct.

This activity demonstrates layering over time, and shows how a core sample provides clues to the past. Divide students into pairs or groups. Each student group will create a simulated sediment bed imbued with pollen grains, then exchange beds with another group. After layering the sediment and pollen in buckets, students take core samples using modified syringes then analyze the samples.

Hints

• In preparation for this activity, call an agricultural field station or the geology department of a local university to find out if any core samples have been taken in your area, and what information was discerned from them.

• Field guides often include natural history, and are available through most local libraries.

Introduction Guide

a. All seed plants produce pollen; pollen contains sperm needed to fertilize flowers. All pines produce pollens, as do flowering plants and grasses, but not mosses and ferns.

b. Pollen is taken to and from plants by wind, bees, flies, butterflies, hummingbirds, bats, moths, human beings, dogs, and other passersby.

Discussion Guide

a. This will depend on your location. Look at trees growing nearby. A good field guide to trees and plants will help students compare local species with those growing elsewhere.

b. There would be less pollen from pine trees, as these grow best in cold climates, and more pollen from tropical plants.

c. Volcanic ash, fish scales, and plankton are other materials scientists study for clues.

Rocky Records

GEOLOGY

Background information

The limestone spires known as stalactites and stalagmites offer a layered record of Earth's past climate. Periods of time when conditions were cold or warm and wet or dry can be identified by the thickness and composition of their layers, then dated by measuring their radioactive isotopes.

Stalactites and stalagmites are collectively called speleothems. They grow in caves as the calcium carbonate (**calcite**, for short) in limestone dissolves in, then separates out of, groundwater. How does calcite first become dissolved in water? Many underground caves are carved out of limestone, a sedimentary rock composed mainly of calcite. Calcite does not usually dissolve well in water, but groundwater descending through layers of soil picks up carbon dioxide from decomposing organic matter. If it absorbs enough CO_2, the water becomes slightly acidic—in fact, it becomes a weak solution of carbonic acid that dissolves calcite as it drips downwards through limestone cave ceilings.

But the calcite does not stay in its dissolved state. As it emerges into the cave, water carrying the calcite encounters air that contains less CO_2 than it does. Gases continually adjust to their surroundings to maintain a state of equilibrium, so some of the water's CO_2 transfers into the surrounding air. This shift in molecular makeup has dramatic repercussions, including the jettisoning—or **precipitation**—of the dissolved calcite. Released from its watery solution, the calcite takes solid form once again, heaping up onto earlier deposits of calcite until a spire begins to form.

The rate at which speleothems grow depends on a number of factors, the most important of which is climate. A wet, warm climate provides plenty of water to carry dissolved minerals. Water and warmth also lead to increased plant growth, which in turn enhances the CO_2 content of the soil. This renders groundwater more acidic, and more prone to dissolve calcite. So it is possible to make inferences about the variability of past climate from the layered structure of a speleothem: thicker layers indicate warmer, wetter conditions.

How do we know when layers in a speleothem formed? Dating of speleothems is conducted by measuring uranium and thorium isotopes deposited by the calcite. Radioactive parent ^{234}U is soluble in groundwater, whereas its radioisotope daughter ^{230}Th is not. Following deposition, uranium decays, producing ^{230}Th at a known rate. So, relative amounts of the thorium and uranium isotopes are a measure of the time elapsed since deposition.

Objective:
To understand the formation of stalactites and stalagmites by creating models from wax and demonstrating the precipitation of crystals from solution.

Materials:

3–4 tapered multicolored candles, ideally composed of concentric circles

3 100-gram blocks of white paraffin

2–3 deep-colored crayons per 100-gram block of paraffin (*e.g.*, 2 dark blue, 2 deep red, 2 bright yellow)

newspapers to protect work area (including the floor)

3 long-handled wooden spoons

3 deep metal pots or coffee cans

3 electric hot plates

sharp serrated knife

colored pencils

Introduction

a. The mechanism involved in speleothem growth is chemical precipitation. Solids can form by a number of different processes. List examples of buildups that occur in the human body, in your home, or in nature. Which involve chemical precipitation of a solid from solution? Which involve physical accumulation? Which involve biological growth?

b. Demonstrate the dissolution and re-precipitation of a solid. Pour warm water through a small strainer filled with rock salt; collect the water that drains out. What do you observe, and how do you explain it? Boil away the collected water until it all evaporates. What remains? Alternatively, mix equal volumes of sand and salt. Place the mixture in a funnel lined with a coffee filter. Pour water through the funnel collecting the solution. What happens to the amount of material in the funnel? Why? Boil the collected solution to precipitate the salt.

Build a Model Speleothem

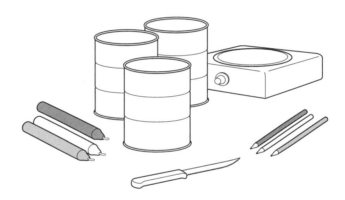

Procedure

a. Melt the white paraffin and crayons together in separate pots or coffee cans over hot plates.

b. Designate crayon colors to represent different climates (*e.g.*, cold and wet, warm and wet, warm and dry, hot and dry). Record which color represents which climate. For example, blue = cold, wet.

c. Holding the candles upright, dip each into one color wax. Repeat several times to give some thickness to the color coating. Move on to other colors, varying the thickness of the layers. Repeated use of colors is encouraged: blue, red, blue, yellow, etc. Even if using already full-sized candles, dip and

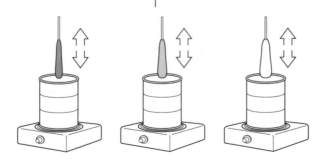

coat candles to view the deposition of layers. Record time spent and number of dips in each color on the chart below.

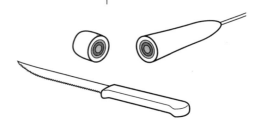

d. When the candle reaches a diameter of approximately two inches, allow it to harden fully (for approximately 12 hours). Then use the knife to slice through the candle horizontally. This will expose various colors in cross-section.

e. Draw the cross-section with colored pencils. Try to make each layer of the colored cross-section to scale. Then number each layer from inside to outside (*i.e.*, from older to younger).

Time Start	Time End	Color	Duration	Number of Dips

Crystal Growth

Procedure

Part 1

a. In 2-liter container, mix 1.5 liters warm water and 10 tablespoons Epsom salts. Stir in additional Epsom salts until the solution is saturated (*i.e.*, no more will dissolve).

b. Pour half of the solution into each 1-liter jar. Place the jars about 10 cm apart on a piece of foil.

c. Soak the string in one of the jars, remove it, then drape it

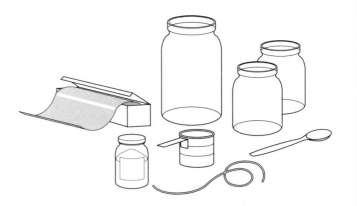

Crystal Growth Part 1 Materials:

aluminum foil to cover work area

one 2-liter container

240-ml measuring cup

1.5 liters warm water

tablespoon or measuring spoon

Epsom salts (hydrated magnesium sulfate, $MgSO_4 \cdot 7H_2O$)

two 1-liter jars

35-cm length of cotton string

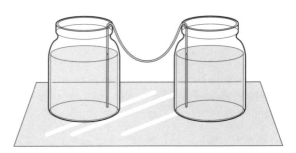

between the jars. Be sure each end of the string is soaking in one of the jars, and that the lowest point in the sag between them is below the level of the solutions in the jars.

d. Leave the jars and string undisturbed for several days. A stalactite will form from the sag in the string.

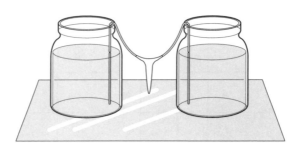

Crystal Growth Part 2 Materials:

4 or 5 charcoal briquettes

2-liter glass bowl

small glass or cup

1 tablespoon household ammonia

2 tablespoons water

1 tablespoon table salt

2 tablespoons laundry bluing

Part 2

a. Place charcoal briquettes in a large glass bowl.

b. Mix ammonia, water, salt, and bluing in a small glass bowl. Pour the mixture over charcoal.

c. Allow the bowl to sit undisturbed for 72 hours. Fluffy, white crystals will form on top of the charcoal and on the sides of the bowl.

Discussion

a. Compare the crystals that grew in Part 1 and Part 2. Consider the physical properties of the crystals (*e.g.*, shape, color, hardness) and the processes by which they formed.

b. Did the two crystals form in the same way?

c. Which demonstration more closely simulates the formation of speleothems and why?

d. How do the formation of speleothems and crystals differ?

e. Based on the climates assigned to each color, write a paragraph describing the history represented by your speleothem. What natural events caused the changes in climate? How long did each period last?

f. Discuss how misleading it is to calculate the passage of time from speleothem thickness. Take this opportunity to discuss the importance of corroborative evidence such as that provided by radioisotopic dating.

Teacher's Guide to the Activity

Time:

Build A Model Speleothem takes 15 minutes to prepare. To conduct the activity takes 50 minutes. Crystal Growth, Part 1 takes 10 minutes to prepare and 15 minutes to conduct, with a break of several days; and Crystal Growth Part 2 takes 15 minutes to prepare and 15 minutes to conduct, with breaks of several days.

In this activity students create speleothems and crystals to demonstrate the way stalactites and stalagmites form. Students build a model speleothem by dipping candles into melted wax of various colors, allowing the wax to dry then slicing through to reveal the many layers. Each color should represent a different climate. In other activities, a stalactite forms as an Epsom salt solution drips down from a soaked string whose ends rest in jars of solution, and crystals form in a bowl of charcoal briquettes coated in a mixture of ammonia, salt, and water.

Hint

• Hot plates and hot wax can cause injury; their use requires careful supervision. Perform this part of Build a Model Speleothem as a demonstration if necessary.

Introduction Guide

a. Examples of buildup include tartar on teeth, wax on furniture, dust and dirt, cholesterol deposits in veins, mineral scales in pipes, salt crust from an ocean swim, algal slime on a pond, mold on leftovers in the refrigerator, wax in the ears, and ice on the insides of a freezer. The mechanism involved in speleothem growth is chemical precipitation.

b. The rock salt grains become smaller when warm water is poured through it because some of the salt has dissolved. When the collected water is boiled away, the salt precipitate remains. The volume of material in the funnel will decrease as salt dissolves and washes from the sand. Point out that not all solids dissolve readily in water.

Discussion Guide

b. Not exactly. In both demonstrations, evaporation of water made the crystals precipitate. This means that a solid formed in a solution. In the case of Epsom salts, the crystals that precipitated are composed of the ingredient we started with: Epsom salts. In the second case, the crystals that formed were not made from one of the original ingredients. We can infer that these crystals are a product of some reaction(s) that took place in solution among the original substances we added.

c. The Epsom salt "stalactite" shows the form of a structure that grows as water drips down. In addition, the end product (a magnesium salt) is the same as the one initially dissolved. Likewise, calcite is the substance that dissolves and re-precipitates in the formation of speleothems. The growth of crystals on charcoal, however, is illustrative of a chemical reaction. Recall that in speleothem formation, chemical reactions are also involved: the reaction of water and carbon dioxide to make carbonic acid, the reaction between calcite and carbonic acid, and finally, the release of carbon dioxide to re-precipitate calcite.

d. Speleothems are useful recorders of past climate because they form so slowly—over many thousand of years, not days. Also, calcite precipitation is controlled predominantly by a release of carbon dioxide gas from groundwater into cave air, not by evaporation as in our demonstrations. Lastly, although calcite is the predominant mineral in speleothems, stalagmites and stalactites also contain trace elements (uranium and thorium) that allow scientists to date them, as well as oxygen isotopes permitting inferences about past temperatures.

e. For example: First there was a stormy, cold period that lasted for about four thousand years. Then plate tectonics brought several continents closer to the equator, and temperatures rose. This was followed by a brief period of cooling because several large volcanoes erupted and the airborne ash blocked the sun for several years. But the long-term warming trend resumed, and has persisted for at least 11 thousand years, to the present day.

f. During colder, drier conditions, speleothem growth rates are reduced; a millenium of cold, dry weather could leave only a thin layer in the rock record.

GEOLOGY

Objective:

To expose various soil samples to sunlight, testing for wetness to show relation between heat energy, evaporation, and erosion.

Digging It

Background information

Severe drought struck this country's Great Plains during the Depression. Seared by the sun and lacking moisture to keep them alive, grasses and other plants normally holding the soil in place withered and died. Left unprotected, topsoil was lifted by strong winds and carried many miles. "I suspect . . . when people along the seaboard of the eastern United States began to taste fresh soil from the plains 2,000 miles away," wrote one observer in 1934, "many of them realized for the first time that somewhere something had gone terribly wrong with the land." The great destruction caused by dust storms introduced this nation to the need to study the interrelationship of heat energy, evaporation, and erosion. Even today, wherever the sun shines steadily, soil surfaces are loose and dry and vegetation sparse or absent unless control measures are applied.

The upper layers of soil (called **topsoil**) are susceptible to erosion because their particles are small and light enough to be easily blown away. This is problematic because topsoil contains many nutrients such as carbon, oxygen, and other materials plants use to make food. If these valuable substances are carried away, plants are deprived of the building blocks they need to survive. Topsoil also holds water, storing it between periods of rain for continual use in photosynthesis. In contrast, erosion leaves behind large, coarse particles that are not only nutrient-poor but relatively non-absorbent, so plants in eroded areas dry out and die more quickly. In general, soil erosion leads to decreased production of the plants that feed the world's population, that take in CO_2, a primary greenhouse gas, and that make O_2, a gas needed for human and animal life to survive.

If it occurs, global warming might contribute to erosion beyond that now caused by wind and water. A rise in worldwide heat energy would have many consequences for this planet, including increased evaporation of water from Earth's surface. Increased evaporation would dry soils, rendering them more susceptible to wind erosion. In fact, wind itself might be affected by global warming. It has been predicted that global warming will cause temperatures at the poles to increase by a greater amount than at the equator. One potential effect of such a change would be dramatic shifts in wind patterns. Elsewhere, increased evaporation of water from the Earth's surface might cause increased cloudiness, more rainfall, and more erosion caused by water. Lastly, if global warming causes polar ice caps to melt and sea levels to rise, flooding would worsen coastal erosion.

In summary, global warming might increase erosion by wind and by water. One chain of consequences may be an overall decrease in plant production, enhancement of the greenhouse effect, and greater global warming.

Introduction to activity

a. Collect samples of the following types of sediments: pebbles or fine gravel (approximately

1 cm in diameter); sand such as beach sand; and topsoil or potting soil. On a sheet of white paper, make three small piles from each sediment. The three piles should be placed side by side, but should not touch. Three students might fan or blow on the piles simultaneously. Which travelled furthest? Why?

b. Collect additional sand or soil from different sites, noting each location where sediments were removed so they can be returned to the same place. At each site, before removing the sand/soil, work it with your fingers. How does it feel? Does it feel different at different depths? Is it dry, crumbly, coarse, moist, spongy, drippy, sandy? Do the samples smell different?

c. Take a walk to observe evidence of wind and water erosion in the neighborhood of your school or home. What are some shared features of the eroded landscapes?

Procedure

a. Prepare the plastic cups by carefully punching 3-4 holes into the bottom of each.

b. Line the bottom of each glass with a piece of paper towel to keep soil from falling out of the holes.

c. Distribute the sediments so that the pebbles, sand, and soil fill four cups, respectively. Place the cups in a plastic tub or sink.

d. Starting with half a cup of water, slowly pour water into each cup until all contents appear thoroughly wet and excess water begins to drain from the cups. (Allow each cup to drain for about five minutes.) Which sediments soaked up the most water?

e. Using spoons, transfer the contents of each cup into a plastic dish and spread the sediments fairly evenly. Group the dishes into four sets of three, so that each group contains one dish with pebbles, one with sand, and one with soil. Weigh each dish on the weighing balance and record the results.

f. Place one group of dishes into a covered box or a dark closet. (This is the "dark" treatment group.) Place one group of dishes on a sunny window sill or directly underneath a bright lamp. (Do not use fluorescent lamps, as they give off little to no heat. This is the "light" treatment group.) Place the third group of dishes near a fan positioned so as to sweep their surface gently. This is the "wind" treatment group. Finally, place the last group of dishes in a second sunny spot near a second fan. This is the "wind & light" treatment group, which should be kept separate from the "light" treatment group.

g. Re-weigh each dish every hour for one day, recording the results. The longer the period of time over which observations are made, the more dramatic they will be. By monitoring the weight change in each dish, we are essentially observing the effects of evaporation. Greatest drops in weight will correspond to the most evaporation.

h. Calculate the change in weight observed for each treatment group by subtracting the final weight measurement you made from the initial weight. Record this.

Materials:
12 identical clear plastic cups

pen or other object with pointed end

1 roll of paper towels

enough pebble or fine gravel, sand and potting soil to fill 4 cups, respectively

plastic tub (or sink) large enough to hold all cups

pitcher of water

measuring cup

one spoon per student in work group

12 identical plastic picnic plates

weighing balance

large box with cover (or a dark closet)

paper

Discussion

a. Which type of sediment (pebbles, sand, or soil) dried out most quickly? Which sediment types showed the biggest change in weight?

b. Which type of sediment had the highest capacity for holding water? Which one showed the smallest change in weight?

c. Which treatment group—dark, light only, wind only, or wind and light—dried the soil fastest? What does this suggest about the effects of natural elements on erosion?

d. What are some ways that farmers could help prevent erosion in their fields?

Teacher's Guide to the Activity

In this activity students expose various soil samples to sunlight to highlight the relationship between heat, evaporation, and erosion.

Introduction Guide

b. If the soil is clumpy, it probably contains clay. If it is loamy (*i.e.*, mouldy), it probably contains a good deal of organic matter, such as dead leaves.

Discussion Guide

a. The pebble/fine gravel should dry out first, followed by the sand and potting soil.

b. The potting soil should have the highest capacity for holding water. The sand might also stay wet for a long time.

c. The combined treatment of wind and light should dry the soil the most quickly, though this depends how intense the light was, how strong the wind blew, and whether or not sampling every hour is often enough to detect a change.

d. Whenever soil is exposed the chances of erosion increase, so many farmers plant substitute vegetation when their crops are not being grown. This means there are always plants in place fighting the effects of wind and water erosion. The plants' roots help to stabilize the soil, and the land is protected from the wind and rain. Planting replacement vegetation also nourishes (*i.e.*, fertilizes) the soil. Farmers might also **terrace** their land, or build step-like fields so rain water will be absorbed by any number of level surfaces instead of running downhill. Finally, farmers also plow the land to divert flows of water around their fields. This is called **contour plowing**.

Time:
This activity takes 15 minutes to prepare; it takes 50–60 minutes to conduct, followed by 5 minutes at one-hour intervals.

METEOROLOGY

Objective:

To measure the changes in evaporation rate of water in different forms at different temperatures.

Going, Going, Gone

Background information

Water exists in three forms called phases: solid, liquid, and gas. We commonly call these phases ice, water, and water vapor, respectively. Heat energy is required to change ice to liquid water and to evaporate liquid water. When the energy from the Sun reaches Earth, some of it is absorbed by the land and the water. The energy absorbed by water molecules in oceans, lakes, and streams causes these molecules to move about more and more quickly. If enough energy is absorbed, some surface water molecules break free from neighboring molecules and rise into the atmosphere as water vapor. This is the process of evaporation.

Phase changes notwithstanding, one of water's crucial aspects is its constancy. Regardless of its appearance, the total amount of water on or above the planet hardly changes. Some water vapor is lost to space and some water becomes "locked" into rocks, but to a surprising extent the quantity of water on Earth is constant. Water might change forms, appearing as ice, liquid water, or as water vapor; however, rain isn't new water. There is virtually no new water. Almost all the water that was, that is, and that will ever be has been present on this planet in one form or another for many millennia. In other words, some chemical processes aside, we drink the water the dinosaurs drank! That's right! Virtually all water has been warmed by the Sun innumerable times. It has evaporated into vapor innumerable times. It has fallen as rain over and over and over again. It has frozen into snow and ice almost as often.

The total amount of water on Earth does not change when the global climate warms or cools, but the fraction of the water that is liquid, solid or gas does change. So does the distribution of the water on the planet. For example, at times when the global climate is cooler, more water might be frozen as polar ice, leaving less liquid water in the oceans. During times when the Earth's atmosphere warms up, the rate of evaporation increases, and so does the amount of water vapor in the atmosphere. This increase in humidity results in more clouds and hence more rain. During colder periods, there is less evaporation and humidity is low, so there are fewer clouds and less precipitation.

Introduction to activity

a. List everyday examples of evaporation. Discuss what sorts of conditions speed up or slow down evaporation. Do clothes hung outside dry faster when the weather is hot or cold, windy or still? Why?

b. Devise strategies to speed the disappearance of a small puddle (2–3 drops) from a piece of waxed paper. Which strategies are predicted to work best? Which do?

c. Discuss the order in which water's changes in phase can occur. Draw a picture showing the three phases of water and connect the phases with arrows to show common transformations. Can vapor and solid be connected?

d. Scientists are working to understand how climate change might affect the rates of natural processes such as evaporation. In their experiments, comparatively normal conditions—generally called **controls**—are compared to novel, or **experimental**, conditions. What would the control condition be for a classroom experiment on evaporation? List good places to simulate the experimental condition of a warm climate.

Procedure

a. Decide on four classroom locations that have different temperatures (*e.g.*, on a windowsill in the sun, on top of a bookshelf near room lights, in a dark corner, in a refrigerator, in a freezer, on the teacher's desk).

b. Label the sixteen cups: four "ice," four "snow," four "water," and four "wet earth."

c. Place one ice cube into each cup marked "ice."

d. Using the plastic bag and hammer, crush one ice cube. Place the resulting "snow" in one cup marked "snow." Repeat this procedure to fill each remaining cup marked "snow." Do not crush more than one ice cube at a time, as volumes must be as close to identical as possible for later comparisons to be legitimate.

e. One at a time, pour one ice-tray section of water (having volume equivalent to one ice cube) into each of the four cups marked "water."

f. Use the tablespoon to place the same amount of soil in each of the cups marked "wet earth," then add one ice-tray section of water to the soil. Mix the soil and water.

g. Group the cups into four sets of four, so that each group contains one cup of ice, one of snow, one of water, and one of wet earth. Place one set of cups in each of the locations simulating control and experimental conditions. Make an initial observation of water level in each cup, and record the experiment's starting time.

h. Once an hour (or at other pre-arranged times), check how much water remains in each cup. Record how long it takes for each cup to dry out completely. Check the soil for moisture by blotting it with the filter paper and looking for signs of wetness.

Discussion

a. Which of the forms of water evaporated first? Why?

b. How might rates of evaporation be speeded?

c. What could happen to evaporation rates in the event of global warming?

Materials
16 identical small plastic cups

marking pen

1 ice tray

8 ice cubes (identical volume, frozen in above tray)

1 small, tough plastic bag

hammer

1 tablespoon or comparably-sized spoon

soil to partially fill 4 small cups

16 coffee filters, cut into small (*i.e.*, 8 x 8 cm) pieces

---

d. How have ice caps at the North and South poles continued to exist in the face of past periods of warming?

e. Besides increased evaporation and resultant precipitation, what other consequences of global warming might arise?

Teacher's Guide to the Activity

By observing and measuring rates of evaporation of water in several forms, students learn the relationship between temperature, form, and evaporation.

Introduction Guide

a. Examples include puddles drying up, clothes drying on a line, and potted plants drying out.

b. Strategies include spreading the water out, placing it in the sun, and fanning it.

c. An arrow drawn from a liquid to vapor represents evaporation, and one pointing the opposite way represents condensation. Arrows between liquid and solid represent freezing and melting. Vapor and solid can be connected. The process of going from ice directly to water vapor is called **sublimation**, but it rarely occurs. It is far more common for ice to melt into water before the water evaporates.

d. Water at room temperature would be the control condition for a classroom experiment on evaporation. A sunny window or heated radiator would be a good place to simulate the experimental condition of a warm climate. A cool climate could be simulated using a dark closet or refrigerator.

Discussion Guide

a. Provided all of the initial volumes of water were truly equal, the liquid water should evaporate first. This is because the snow and ice must melt to liquid before they can evaporate.

b. Wind blowing on water's surface speeds evaporation by moving away vapor directly above the surface. This frees space for more water vapor to enter the air.

c. Evaporation rates are predicted to increase with global warming.

d. Precipitation is needed to maintain thick ice packs. For fields of ice to continue to exist, there must have been enough of a balance between warm conditions that allow evaporation and cool ones that prevent excessive melting of the ice.

e. More water vapor in the atmosphere might lead to more clouds. Clouds can have two opposing effects with regard to regulating Earth's temperature. First, they reflect some of the Sun's energy, preventing it from reaching Earth's surface. This cooling effect tends to prevent further increases in temperature, and keeps evaporation rates in check. The second effect of water vapor in the atmosphere is that it traps heat, like a blanket around Earth, causing global temperatures to rise further. The first effect is an example of negative feedback; the second is an example of positive feedback. Another example of positive feedback concerns the effects of global warming on Earth's ice and snow. Surface snow and ice reflect about 10 percent of incoming solar radiation, slowing the warming of Earth. As the temperature rises, however, more of this ice will melt. Hence more and more heat will be absorbed by the exposed land surfaces, raising temperatures still further.

Time:
This activity takes 15 minutes to prepare; to conduct, it takes 50 minutes, then 10 minutes over one-hour intervals.

METEOROLOGY

Objective:
To understand how the distribution of water creates climate zones on Earth.

Making Weather

Background information

San Francisco, Kansas City, and Washington, D.C. These cities are located in the same temperate zone of the Northern Hemisphere, yet summer highs are usually in the low 20's in San Francisco, the low 30's in Kansas City, and the mid 20's in Washington D.C. In winter, lows are in the low teens, nearly minus 10, and minus 5, respectively. With respect to rainfall, San Francisco receives an average of 50 cm, while Kansas City is wetter, at 90 cm, and Washington, D.C. is wetter still at 100 cm. Why do these cities have such dissimilar climates? Currents and winds distribute heat and water across the United States. It is variability in these distribution patterns that creates quite different climates.

San Francisco is a coastal city. Not far off its shore ocean temperatures are a chilly 10-15°C thanks to the California Current, which carries cold water south from the Gulf of Alaska. Prevailing winds from the west sweep San Francisco with air and moisture cooled by the ocean. In contrast, the air creating Kansas City's weather takes its temperatures and humidity from the land over which it travels. Kansas City is so far inland that winds reaching it have blown westward, over more than 2,500 kilometers of land that heats and cools much more quickly than the ocean does. Given this, temperatures in Kansas City fall lower and rise higher than temperatures in San Francisco. Prevailing winds usually blow from the west to Washington, D.C., too, carrying air brought to land temperatures over days or weeks of travel. But Washington, D.C. is also positioned near the Gulf Stream, which brings warm water north from the Caribbean and Gulf of Mexico. Temperatures of ocean water off Washington, D.C. are about 15-25°C, so air tends to be more moist and its temperatures in a smaller range than Kansas City's.

When we speak of weather and climate, then, we are referring to the transfer of water and heat from one place to another. These transfers often require that water change phases. For example, liquid water in the sea or in a lake might evaporate, or change from liquid to gas. Individual molecules of water in the atmosphere might then condense, coming together again to form liquid water. Droplets of water might then fall back to Earth's surface as rain, or freeze to form ice crystals if the air temperature is low enough. The ice might melt into liquid water again, and so on. However, these changes in phase occur unevenly over the surface of the planet. Water that evaporates from one location might fall as rain in a very distant place. The patterns by which water is redistributed help create vastly different climate zones. In the event of global climate change, these patterns might change in ways that will affect life all over the planet.

A Recipe Using Water

Introduction to activity

a. Use a newspaper to locate and compare today's temperature highs and lows in San Francisco, Kansas City, and Washington, D.C. Identify three other cities positioned on exactly the same latitude. How similar or dissimilar are these cities' weather and climate?

b. Sketch a thermometer, marking on it the freezing and boiling points of water in degrees Fahrenheit and Celsius.

c. Make a list of where you have observed water in each of its three phases.

Procedure

a. Observe the empty jar.

b. Fill the jar part way with hot water and cover it with its lid.

c. Observe the steam that rises from the water in the jar.

d. After several minutes, open the jar carefully. Notice that the inside surface of the lid is wet where the rising vapor has condensed upon it.

e. Reseal the jar and place it in the freezer several hours or overnight.

f. Observe the ice that has formed inside the jar.

g. Set the jar aside at room temperature, and observe that the ice melts into liquid water.

Discussion

a. List the changes in phase that took place inside the jar in chronological order. Decide if each change in phase was the result of cooling or heating.

b. Pair each change in phase that occurred in the jar with a similar transformation in nature.

c. It is probably clear that in a closed jar we can make the water change phases indefinitely, simply by heating and cooling the jar. What would happen if we left the jar open?

d. Why doesn't Earth, like the open jar, dry out? Why doesn't all the water escape into space?

e. What is different about the hydrologic cycle inside the jar and the global-scale hydrologic cycle?

f. If planet Earth loses almost no water, simply recycling it again and again in different phases, why should we worry about conserving water? How does this demonstration relate to the subject of global climate change?

Materials

clear glass jar with dry, tight-fitting lid

hot water

freezer

national weather report from a newspaper

Teacher's Guide to the Activity

Time:

This activity takes 10 minutes to prepare; it takes 10 minutes to conduct, with a break of several hours or overnight, then another 10 minutes.

The presence or absence of moisture and changes in temperature create distinct climate zones. In this activity, students observe phase changes in water due to heating and cooling and are asked to relate these changes to climate zones.

Introduction Guide

c. Liquid water is found in puddles, streams, bathtubs, and combined with other substances in beverages we drink. Solid water is found in icicles, ice skating rinks, ice cubes, the freezer, and igloos. Water vapor is trickier to observe because it's virtually invisible. Examples of water vapor include high humidity felt on a muggy summer day, and steam (actually tiny drops of liquid water on their way to becoming a gas) rising from boiling liquid.

Discussion Guide

a. The phase changes include:
1. Liquid water turned into water vapor.
2. Water vapor condensed into liquid water.
3. Liquid water froze into solid ice.
4. Solid ice melted into liquid water.

Using the answers above, 1 and 4 are examples of heating while 2 and 3 are examples of cooling.

b. Examples include:
1. The sun shines on a lake, warming the water, and some of the lake water evaporates.
2. As the water vapor rises and cools, it condenses to form water droplets in clouds. Rain falls from the clouds.
3. Winter begins and temperatures fall, so the lake begins to freeze.
4. In the spring, temperatures rise, and the ice melts.

c. Eventually all of the water would evaporate and escape from the jar.

d. As water vapor rises from Earth's surface it ascends into cooler regions of the atmosphere and loses some of its heat energy. Its molecules begin to slow, and gradually bond together again. If the molecular clusters grow large and heavy enough, gravity pulls them down in the form of rain, snow, sleet or hail. So the temperature gradient in the atmosphere and the force of gravity keep Earth's water on the planet.

e. There are many differences. Here are two examples:
1. Ocean water contains many more salts, minerals and organic matter than drinking water. What happens when saltwater evaporates? Find out by devising a simple test.

2. Throughout all changes in phase, the water always ended up in the same place it started—in the jar. On Earth, water that evaporates and condenses usually falls as rain or snow somewhere other than where it started out. For instance, what happens when water starts as solid ice? Where does snow on a mountain top go when spring or summer arrives? Consider contacting a ski resort to inquire what special arrangements must be made to cope with large amounts of melting snow and ice.

Other differences between weather in a jar and weather in the world include plants' and animals' uses of water. Human beings drain underground water tables, dam rivers, and import water to arid areas via pipeline. Winds and currents also affect the water cycle.

f. The total amount of water on the planet is not our chief concern. Rather, we worry about the *quality* and *distribution* of the water. Almost 97 percent of all the water on Earth is saltwater. Of the remaining 3 percent, almost all is ice. Only a tiny fraction can be used by animals (including human beings) and plants. So it is important that we utilize our fresh water wisely and not contaminate it.

It is also important to think about the location of fresh water on Earth. All life requires water for survival, and the distribution of life on Earth is largely dictated by the distribution of fresh water. Plants and animals tend to live near rivers, lakes, and groundwater reservoirs. But as human populations continue to grow, the distribution of people appears to be less dependent on the location of natural water sources. As an example, consider cities situated in the desert. In areas where water must be imported, there is a need to consider the consequences of depending on water transported over long distances.

In the future, global climate change could cause the distribution of the total planetary water resources to change. For example, a larger fraction of fresh water may melt from glacial ice. Regional changes in climate could cause areas now rich in water to become drier. A larger fraction of Earth's water might become seawater rather than fresh water. Changes such as these would increase the need for global-scale planning and conservation.

METEOROLOGY

Objective:
To identify, measure, and average microclimatic temperatures.

It All Adds Up

Background information

Have you ever noticed how much cooler it is in the shade than in direct sunlight? Or how much hotter it feels to stand on pavement as opposed to a grassy patch of land? Temperature differences within a small area are indications of **microclimates**: very small-scale climate conditions. The following are a few examples of microclimatic variation:

- dense, cold air sinking into the bottom of a valley can make the valley floor 20°C colder than a slope only 100 meters higher;

- winter sunshine can heat the south-facing side of a tree (and the habitable cracks and crevices within it) to as high as 30°C, while the temperature only a few centimeters away from the tree is below freezing;

- the air temperature in a corn or wheat field can vary by 10°C from the soil to the top of the canopy.

Frogs, beetles, and other small animals experience temperature changes on even smaller scales (*e.g.*, pockets of coolness formed by crevices in tree bark, the shade of a leaf, or moist soil beneath a rock). Such small-scale temperature variations might seem unimportant, but they help set the distribution of organisms. Scientists seeking evidence for the influence of temperature on living organisms do not confine themselves to obtaining global or regional temperatures. In fact, average temperatures collected over large areas or over long periods of time offer only rough explanations for specific animals' distribution and abundance.

Air temperature varies from one location to another (spatially) and from one time to another (temporally). For example, there are large variations in temperature with latitude (it is warmer at the equator and cooler at the poles) and over seasons (it is warmer in summer and cooler in winter). Given these variations, how do meteorologists know what the planet's temperature is? Average global temperature can be determined by dividing the globe into a grid and averaging temperatures collected from weather stations in each cell of the grid. Local temperatures reported on the evening news and in daily newspapers are determined much the same way, but on a regional scale. The same principle of averaging temperatures to calculate a single temperature for an area can be applied to the classroom and the school yard.

Introduction to activity

a. Discuss the difference between the following pairs of terms: climate vs. microclimate; spatial vs. temporal variation; climate vs. weather. Give examples of each.

b. Temperature is a good example of an important environmental condition that might vary on a microclimatic scale. List other environmental factors which might vary on a relatively small, or microclimatic, scale.

c. Discuss the ways in which changes in environmental conditions can affect temperature.

Procedure

a. Working individually or in small groups, identify and draw a small-scale map of an area to be sampled (*e.g.*, the classroom, playground, park, backyard).

b. Divide the map into a grid and identify potential locations for microclimatic differences (*e.g.*, under shaded rocks, on the sunny side of a building, on blacktop vs. a grass field).

c. Take thermometers to different locations identified on the map. Record the air temperature in these locations, making sure enough time is allowed for thermometers to acclimate to their surroundings (approximately 5-10 minutes).

d. Record the data on the appropriate grid of your map, and calculate an average of all the data you have collected.

Discussion

a. Compare the different temperatures on the map. Were temperatures relatively similar at all locations, or were there large variations? Is the average temperature closer to the maximum or minimum temperature recorded?

b. Repeat the experiment on a day when the weather is quite different, and compare the results.

c. Can you identify why some locations are characterized by warmer or cooler temperatures than others?

Materials:
large white piece of paper or chalkboard

pencils and erasers, or chalk

1 thermometer per student

Teacher's Guide to the Activity

Time:

This activity takes 10 minutes to prepare and 60–80 minutes to conduct.

This activity introduces microclimates. Terms are introduced in pairs: climate and microclimate; spatial and temporal variation; and climate as opposed to weather.

Hint

• Have students compare the mean temperature with the median temperature.

Introduction Guide

a. Definitions of terms are as follows:

> *climate:* average long-term weather conditions in any particular region.

> *microclimate:* the average weather conditions, or climate, of a small, specific location within a particular region.

> *spatial variation:* differences between locations.

> *temporal variation:* differences from one time to another.

> *weather:* atmospheric conditions at one point in time.

> *moisture:* liquid diffused or condensed in relatively small quantity.

> *wind stress:* cooling caused by wind carrying heat away from a surface, just as blowing on hot liquids lowers their temperature.

> *insolation:* warming caused by direct sunlight. While at the beach, if you suddenly feel cooler when a cloud blocks the Sun, you experience a drop in the Sun's insolation. Less of the Sun's energy reaches you because the clouds have blocked some of its radiation.

b. The environmental factors which might vary include moisture, wind stress, and insolation—the amount of sunlight vs. shade.

Discussion Guide

a. Answers will vary. These comparisons will help students see how an average temperature for a region might be misleading if one observes only a specific spot in that region.

c. Variations might be due to wind effects, evaporation of moisture, or shielding from the Sun's rays.

Flour Shower

PHYSICS

Background information

Have you ever seen rays of sunshine streaming into a room through a window? How about the beams cast by car headlights into the fog at night? Have you noticed the beam of light that goes from a projector to a screen in a darkened theater?

These beams of light illuminate tiny particles in the air: pollen, mold spores, bits of skin and hair, sand, ash, automobile exhaust, and industrial emissions. More exactly, the light encounters and is deflected by the particles, and some portion of the light is blocked.

What effect do atmospheric particles have on Earth's climate? They slow or block light and heat passing through the atmosphere. The result could be either cooling or heating of the Earth.

On the one hand, less of the Sun's energy reaches Earth if it is blocked by particles. This could serve to *cool* Earth's surface, altering many biological, chemical, and physical processes. For instance, scientists have measured slightly cooler temperatures on Earth following large volcanic eruptions that forced tons of fine ash high into the upper layers of the atmosphere. The ash can stay suspended—and spread—for a month or more, thereby influencing global climate. Longer-lasting and more widely dispersed particles are made up of tiny droplets of sulfuric acid formed from volcanoes' emissions of sulfur dioxide.

On the other hand, atmospheric particles also block and absorb energy traveling in the opposite direction: from Earth back into space. It is thought that atmospheric particles can trap heat close to Earth, contributing to global warming caused primarily by greenhouse gases such as water vapor, carbon dioxide, and methane. The net effect in this case would be to *warm* the Earth, again affecting many physical and life processes on this planet.

It is not yet clear to scientists whether the net effect of particles in the atmosphere will be to warm or cool the planet, though present evidence points to a net cooling. Airborne particles in sunlight's path alter energy transmission in one direction, another, or both, so it is not known whether the ultimate impact of continuously-added particles will be a warming or cooling one. However, we can predict that anthropogenic (human-caused) addition of large quantities of particles to the atmosphere will almost certainly have some effect on global climate.

Introduction to activity

a. Make a list of naturally occurring and human-created particulate matter. Which can be captured or sampled, and where? Where do you think the dust in your house comes from? Estimate the proportions that are naturally occurring and human-caused.

b. Use strips of transparent double-sided tape to collect (press on and pick up) examples of settled particulate matter. Display and caption these on index cards.

c. List processes that remove particles from the air.

Objective:
To demonstrate the filtering effects of airborne particles with respect to heat and light.

Materials:

cardboard box with open top

scissors

writing paper

sheet of waxed paper to catch flour

tape (transparent or masking)

flashlight with bright beam

flour sifter or fine metal strainer

flour

cup to scoop flour from bag

2 identical thermometers

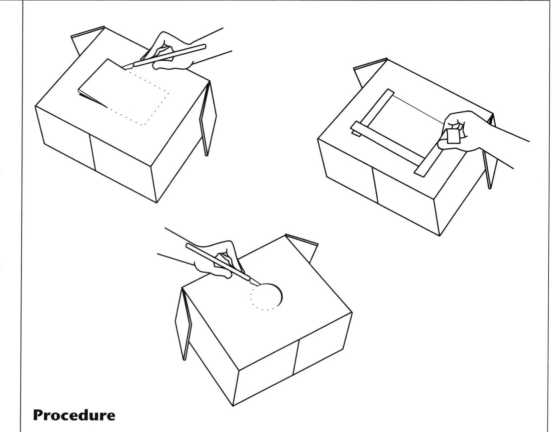

Procedure

a. Prepare the cardboard box by cutting a "window" in one side and taping the paper in place to cover the opening. In the opposite side, cut a hole just big enough to mount the flashlight handle securely. The flashlight should be positioned so as to shine across the box onto the papered window. Position the waxed paper over the bottom of the box to catch as much flour as possible.

b. Turn the flashlight on. Observe the light that shines through from the outside of the papered window.

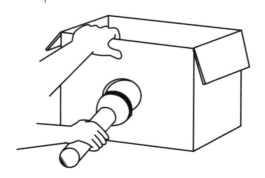

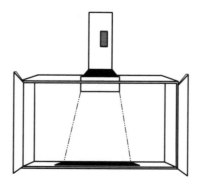

c. Begin to sift flour into the box from above, so that the flour rains through the beam of light.

d. Observe that the intensity of the light seen from outside the papered window decreases because falling flour screens some of the light.

e. Leave the flashlight on for at least 10 minutes, then compare the temperatures inside and outside the box.

Discussion

a. Why do we observe from outside the papered window rather than looking inside the box?

b. What happens to the light that is encountered by the particles, the paper, and the box?

c. What implications does this demonstration have for Earth?

Teacher's Guide to the Activity

Time:

This activity takes 30 minutes to prepare and 50–60 minutes to conduct.

Sifting flour so it falls through a beam of light, students demonstrate the filtering effects of airborne particles with respect to heat and light.

Introduction Guide

a. Naturally occurring particulate matter includes dust, water droplets, ice crystals, volcanic ash, pollen, mold spores, and animal dander. Human-created matter includes bits of skin, hair, automobile exhaust, and industrial emissions. Particulate matter can be captured from inside a car tail pipe, fireplace doors and on a windowsill or a stop sign at a busy intersection.

c. Processes that remove particles from the air include gravity, rain, vacuum cleaners, and a number of filters (*e.g.,* eyelashes and lungs).

Discussion Guide

a. It could be harmful to your eyes to stare directly into light. Also, when gazing at the surface onto which the light shines, we are actually seeing light from two sources: light reflected from that surface and light that is scattered from the particles in the air. Our brain can't sort out which light is which, so it's difficult to detect the effect of the particles on light transmission. From outside the papered window, we see only light scattered by the particles.

b. The light that is blocked from your view does not disappear; it is absorbed in the form of heat. Light is one form of energy. Heat is another. Light energy that is absorbed is converted to heat. You can verify this by measuring the temperature of the air inside and outside the box. Even after the light has been turned off for several minutes, energy absorbed by the box will warm the air inside it. The principle being demonstrated is that any object in the path of the light—a cardboard box, a shower of flour, dust, water droplets, ice crystals, volcanic ash, window shades, or a stone wall—will alter energy transmissions.

c. We have shown that airborne particles can change the amount of light and heat that travel from the Sun to Earth. If these particles continue to be added to the atmosphere, less light and heat might reach Earth's surface, changing many biological, chemical and physical processes. Alternately, the atmosphere might warm as the suspended particles absorb energy and heat up the air around them. This too might alter many natural processes on Earth, including the hydrologic cycle and photosynthesis.

The Reasons for Seasons

PHYSICS

Background information

Most of us are aware that Earth rotates around the Sun. The motion of Earth relative to the Sun is more complex than that of a smaller sphere simply circling around a larger one, however. In fact, several characteristics of Earth's movement in space are important determinants of life. First, the path, or **orbit**, that the Earth follows as it moves around the Sun is not a simple circle. Instead, it is an **ellipse**, or a slightly squashed circle. Second, in addition to orbiting around the Sun, the Earth also spins (rotates) on its own axis, making one complete rotation in a 24-hour period. Think of this axis as an imaginary line through Earth from the North Pole to the South Pole. Third, Earth's axis is tilted with respect to the plane in which it orbits. These three observations explain why we experience night and day; why the relative lengths of day and night vary from place to place and from time to time; and why we have seasons on Earth. Because planetary processes depend on the Sun's energy, variations in the amount of solar energy that reach this planet are largely responsible for the distribution of life on Earth.

As Earth makes its yearly elliptical orbit around the Sun, it comes closest to the Sun two times: in spring and fall. The beginning of these two seasons, in late March and late September, are marked by equinoxes. An equinox occurs when day and night are equal lengths: 12 hours each. The tilt of the Earth's axis brings the Northern Hemisphere closer to the Sun and causes it to experience spring in late March. The Southern Hemisphere, in contrast, is farther from the Sun, and experiences the season of fall. The reverse is true in late September. Winter and summer occur on Earth when it is farthest away from the Sun, on the extreme ends of the ellipse. Along the equator, the imaginary line that encircles the Earth's middle, there are no seasons.

Introduction to activity

a. Use paired small and large balls (*e.g.*, a ping-pong and beach ball, a grape and grapefruit) to represent Earth and the Sun. If possible, mark the equator on the small ball, and label the Northern and Southern hemispheres. Slide a bamboo barbecue skewer through the small ball to represent the Earth's axis. Move the small ball to simulate Earth's motion. Distinguish between rotation and orbit. Be sure that Earth's axis is tilted, and that Earth rotates around its axis rapidly relative to its elliptical orbit around the Sun. Refer to the illustration below. Notice the position of the Earth relative to the Sun for each season.

b. On the board or on graph paper, draw a time line representing one year on Earth. For the Northern Hemisphere, mark the onset of the four seasons, the solstices, and the equinoxes. Mark off the length of time it takes Earth to rotate once around its own axis and to travel once around the Sun.

Objective:

To demonstrate how Earth's shape and tilt cause it to be heated unevenly by the Sun.

Materials:

globe

flashlight

books (for resting flashlight on)

black construction paper

tape

blank paper

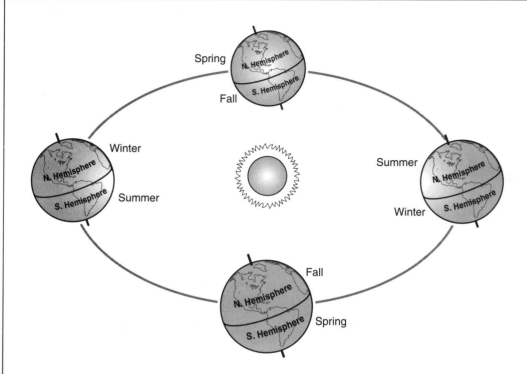

c. Identify possible relationships between the position of the Earth, the length of days and nights, and average temperatures. When are days the longest? When are they shortest? Is there a general relationship between length of day and average temperature? Is it generally warmer when the days are longer or when the days are shorter?

Procedure

a. Place a globe on a table so that the North Pole is pointed directly up.

b. Tape a piece of black construction paper around the end of the flashlight so that it extends several inches past the end of the flashlight. (This helps to narrow the beam of the flashlight.)

c. Turn off the lights in the room, and place the flashlight on a stack of books far enough away from the globe so that half of it is illuminated, and high enough so that the center of the beam is on the equator.

d. Observe whether the Northern Hemisphere or the Southern Hemisphere is receiving the most light.

e. Slowly spin the globe once. Observe whether the amount of light in each hemisphere stays constant while the globe is spinning.

f. Discuss what time of year this should represent.

g. Next, tilt the globe down so that Hawaii is in the middle of the beam.

h. Now which hemisphere is receiving more light?

i. Spin the globe slowly to observe one daily cycle. Does the amount of light in each hemisphere stay constant while the globe is spinning? Which spends more time in the light, Alaska or Antarctica? Discuss why this so. Why do all points in the Northern Hemisphere receive more light than dark during one rotation of the globe? Why does the Northern Hemisphere receive more light during one rotation of the globe than the Southern Hemisphere?

j. What time of year is this?

k. Move the globe such that the equator is in the middle of the beam again. Repeat step i.

l. Move the globe so that the city of Rio de Janeiro, Brazil, is in the middle of the beam. Repeat step h.

Discussion

a. During a solstice, how does day length vary as you move from one pole to the equator, and then to the other pole?

b. On a blank piece of paper, draw the movement of the Earth around the Sun, showing Earth's tilt. Label the various seasons.

c. If the longest day of the year in the Northern Hemisphere is at the end of June, why do the warmest days usually occur one to two months later?

d. How would increasing Earth's tilt affect climate?

e. How are Earth's shape and tilt implicated in the possibility of global climate change?

Teacher's Guide to the Activity

Time:

This activity takes 10 minutes to prepare and 45 minutes to conduct.

This activity models the relationship between the position of the Earth relative to the Sun and the seasons.

Hint

• Discuss students' observations on seasons to lay a foundation for the lesson.

Introduction Guide

a. The graphic below shows the position of the Earth during various seasons.

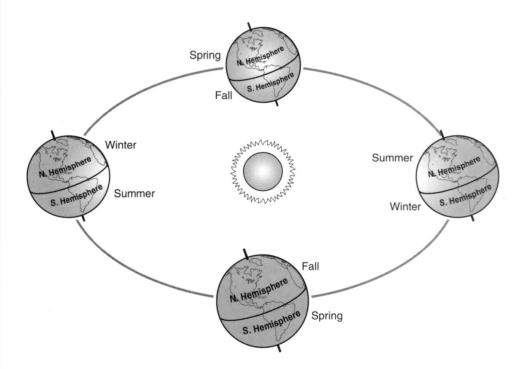

Procedure Guide

d. The two hemispheres should be receiving the same amount of light.

e. The amount of light will not change.

f. This should represent late March or late September, depending on which way the Sun—the flashlight—moves relative to the globe.

h. The Northern Hemisphere is receiving more light.

i. The amount of light will stay constant. Alaska will be nearly always lit and Antarctica will be nearly always dark. More than half the light is falling on the Northern Hemisphere, although the Northern Hemisphere occupies only half of the Earth. Given this, lands north of the Equator spend more time in the light than in the darkness. This is not true of lands south of the Equator.

j. It is late June: summertime in the Northern Hemisphere.

k. This represents late September.

l. This represents late December, or winter in the Northern Hemisphere.

Discussion Guide

a. During summer in the Northern Hemisphere, the day is very long near the North Pole. (Scandinavia and Alaska, for instance, receive sunlight almost 24 hours a day). Toward the equator, daylight becomes progressively reduced, although it is still longer than the night. At the equator, there is always equal day and night (12 hours of each). On the other side of the equator, the length of the day is shorter than the night; particularly near the South Pole. During winter, the situation is exactly reversed. The same fraction of Earth is always illuminated (50 percent), but when Earth is tilted, the hemisphere nearest the Sun receives more than 50 percent of the light.

c. Because Earth's land and water naturally retain heat, it takes some time for further heating to take effect. Likewise, it takes Earth some time to cool down, so the coldest days of the year are usually in January or February, several months after the winter solstice. Given the different heat retaining properties of land and water (water retains heat better then land), the actual warmest and coldest days of the year depend on geographic location.

d. Seasonal differences would become greater if the amount of light received by one hemisphere were to increase. In June, for example, the Northern Hemisphere would receive more light than now and the Southern Hemisphere less. The converse would be true the following December. Thus, seasonal differences both between and within hemispheres would be greater. In actuality, the angle of tilt does vary slightly over long periods of time, and this might have caused differences in climates in the past.

e. Rising temperatures at the equator will almost certainly cause greater warming of tropical oceans than is now the case, giving rise to increased evaporation and clouds. Polar regions might experience even stronger effects. For every one degree of rising temperature at the equator, the Arctic is predicted to heat four degrees. Melting ice packs might cause sea levels to rise and coastal areas to flood, but this would occur over the course of many years.

PLANT BIOLOGY

Plant Power

Objective:

To demonstrate that plants produce oxygen.

Background information

Although most of us have registered environmentalists' pleas to save rainforests and to plant trees, it might not be clear that these actions are important in the face of rising levels of carbon dioxide gas in the atmosphere. The molecule carbon dioxide (CO_2) is one of a number of compounds called **greenhouse gases.** This name comes from the ability of certain gases to trap heat in the atmosphere, warming the Earth. Rising levels of CO_2 and other greenhouse gases, however, could cause temperatures on Earth to increase excessively. Indeed, just such a buildup of CO_2 has been recorded over the past century, caused in large part by the burning of fossil fuels and widespread deforestation.

Plants lessen the effects of human-produced CO_2 in the atmosphere. In producing food for themselves, plants take up carbon dioxide and release oxygen (O_2). This seemingly simple activity, carried out by all plants living on land and in the sea, lowers levels of CO_2 in the atmosphere and provides oxygen for animals (like humans) to breathe. This process is called **photosynthesis. Respiration** is the reverse of photosynthesis. In respiring, plants combine food and oxygen to release the energy stored in these compounds' chemical bonds. Water and carbon dioxide result. This process is carried out by virtually all organisms, plants and animals alike, and is a form of burning, or **oxidation.** Other common examples of oxidation are the flames of campfires, the ignition of car engines, and the rusting of metals, all of which occur only in the presence of oxygen.

Introduction to activity

a. List other reactions that require energy.

b. Discuss your experiences in taking care of house plants. Why is it important to water plants? Why is it important that plants get light?

c. Draw a box diagram showing the cycling of materials between photosynthesis and respiration. Discuss how these occur in nature.

Procedure

a. Prepare the jars by placing 2 votive candles in each jar. Place the plants in one jar. (Make sure the plant isn't so large that it prevents you from closing the jar. If the plants are too tall, pinch off some of the topmost leaves or fronds.)

b. Place all jars in well-lit locations so enough light is available for continual photosynthesis.

c. Carefully light all candles with long matches as simultaneously as possible. Secure lids onto the jars immediately after the candles are lit.

d. Time how long it takes the candles to burn out.

e. Record the rates at which the candles extinguish. Calculate the average duration of candles burned in jars having plants, and compare it to that of candles burned without plants.

Discussion

a. Discuss the processes that are taking place inside the jar. Focus on the means by which plants produce oxygen, but generate alternate explanations of—and possible tests for—the plants' effect on the candles.

b. Discuss the various forms of combustion. How does combustion in fire differ from combustion in animals?

c. Why was it important that the plants in this study were well-watered and well-lit?

d. Discuss the effects of widespread deforestation.

e. Grow plants in the classroom, or organize a planting project around school grounds, to help combat deforestation and to experience plant power in action.

Materials:
2 identical wide-mouthed one gallon glass jars

2 identical lids fitting the jars, made from material that will not melt

several small, sturdy, well-watered plants

4 small votive candles

long matches

stopwatch or clock

Time:

This activity takes 10 minutes to prepare and 50 to 60 minutes to conduct.

Teacher's Guide to the Activity

Students demonstrate that plants release oxygen by placing lit candles in two jars, one with plants, the other empty. Since the candle flames require a continued source of energy in the form of oxygen, without it—as in the case of the candle in the empty jar—the flame goes out.

Introduction Guide

a. Toasting bread requires electrical energy; running a furnace to heat bath water depends on burning oil or natural gas; riding a bicycle takes muscle power derived from food that has been eaten and burned up; sailing a boat takes wind energy.

b. Water and light are needed for photosynthesis. Light energy is used to convert carbon dioxide and water into glucose and oxygen.

c. See diagram below.

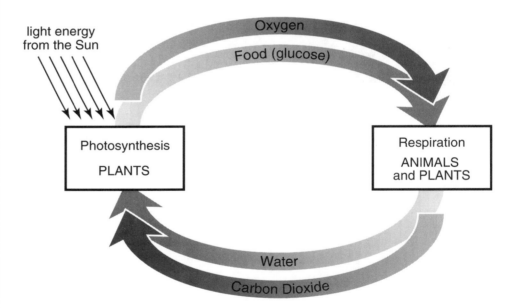

Discussion Guide

a. Plants generate oxygen through photosynthesis. The burning candles represent respiration by animals. Animals consume oxygen when they respire, just as the burning candles use up O_2 in the air. Respiration, like other combustive processes, removes oxygen from the air.

b. In most fires, one large reaction releases large amounts of energy in relatively uncontrolled fashion. In respiration by organisms, innumerable individual reactions occur. Each releases small amounts of energy in much more controlled fashion.

c. Plants need water and light for photosynthesis. To demonstrate this, repeat the experiment in the dark (except for the small amount of light generated by the candles) and/or with plants that have not been watered for some time. To make legitimate comparisons, remember to duplicate all the initial experiment's other procedures.

d. Removal of trees that use carbon dioxide and produce oxygen could allow levels of heat-trapping CO_2 in the atmosphere to rise. Other important effects include loss of species biodiversity and the eradication of indigenous peoples' homelands and natural resources.

e. Here is an example:

student work groups	with plants	without plants
group 1	5.1 min.	2.0 min.
group 2	3.6 min.	1.7 min.
group 3	4.2 min.	2.1 min.
group 4	3.9 min.	1.5 min.
group 5	5.7 min.	0.9 min.
average	4.5 min.	1.6 min.

PLANT BIOLOGY

Leafing Through the Past

Objective:
To demonstrate how scientists decipher clues about past climate from tree rings and leaf stomata.

Background information

Have you ever noticed the patterns in wood grains? Examine nearby furniture or any other object made of wood. Some surfaces show straight lines. In others, the pattern is swirled. If you have ever seen a piece of wood sliced horizontally from the trunk of a tree, you have probably seen a pattern of concentric rings much like that in the surface of an onion cut horizontally. Each ring is really the cross section of a layer of wood the tree added to its trunk each year. If you look closely, you'll notice that the rings have different widths. Wider rings indicate that the tree grew fast; narrower rings signify a year of slower growth. Why do growth rates vary? Growth depends on changing factors, such as climate. Most trees grow faster during cooler, wetter years than they do when conditions are hot and dry, so scientists can use the relationship between ring width and annual precipitation to recreate climate history over the lifetime of the tree. Since many trees live for a long time—sometimes hundreds of years—the climate record preserved in very old trees extends back over many decades. (However, other factors besides climate cause trees to grow faster or more slowly. Injuries caused by fire or insect pests, or crowding by other trees, can slow growth and lead to narrower rings. Interpretation of the tree ring record must be conducted with care.)

Just as tree trunks preserve a record of past precipitation, tree and other plants' leaves offer clues to the composition of past atmospheres. Why are scientists interested in the composition of ancient air? There is a link between the amount of carbon dioxide gas (CO_2) in the atmosphere and the average global temperature. For reasons scientists don't fully understand, more CO_2 was present in the air during times when the climate was warmer. During cooler periods, there was less CO_2. How are samples of the atmosphere from a million years ago obtained? Tiny amounts of ancient air are available in bubbles trapped in glacial ice, but scientists have been able to locate and sample only relatively young bubbles dating back approximately 250,000 years. To learn how much CO_2 was present before then, clues must be sought elsewhere. Fossils of leaves provide some such clues.

On the underside of leaves are tiny pores called **stomata**. The stomata are openings through which gases can enter and leave. In times of warm climate, when the atmosphere is rich in CO_2, plants need relatively few pores to take in all the CO_2 they need, so their leaves develop comparatively few stomata. In times of cold climate, more stomata are needed because less CO_2 is readily available. Given this, scientists can determine from the number of stomata on fossil leaves whether plants lived in warm or cool times. Fossils of leaves as old as 10 million years have been studied to learn about climatic conditions that existed when those leaves grew on living plants.

Examining Tree Rings

Introduction to activity

a. Create a model tree trunk using large flour tortillas and peanut butter. Spread peanut butter between two or three tortillas and roll them together jelly roll-fashion. Starting at one end and moving toward the other, cut the tortillas in several ways: once perpendicular to the roll, again at an oblique angle, and again parallel to the roll. What differences are observable? In this instance, how do layers' orientation vary? What differences in measurements and interpretations are suggested? Scientists try not to cut trees down when studying them. Likewise, at what angles would you make tortilla tree-trunk cores to secure the fullest display of information?

b. Collect samples of wood in which the grain is visible. Try to find pieces that were cut on various angles across and along the trunk of a tree. Discuss why you can see rings in some of the pieces, but only a straight grain in others. Again, what differences in interpretation are suggested by the wood grains' angle and orientation?

Procedure

a. Divide into 4 groups per trunk slice.

b. Using the ruler as a straight edge, draw a cross on the surface of the wood as though cutting a pie into four equal wedges. The marks should look like four spokes of a wheel extending from the center to the edge. Number the spokes 1, 2, 3, and 4 to correspond to the number of student work groups.

c. Count the rings along spoke number 1, from the center to the edge. Record this number, which tells the age of the tree in years.

d. Use the ruler to measure the width of the rings starting at the center. Record the width of each ring in millimeters.

e. Graph the ring-width data on graph paper. (This can be done while fellow group members are still measuring.) Label the X-axis "Age in years." Number the X-axis in one-year increments, left to right, up to the number of years you've agreed to count. Label the Y-axis "Ring width in millimeters" and choose an appropriate scale for tick marks. Numbers will increase in value as you go up the Y-axis. Make sure the range of numbers on the Y-axis will include both the widest and narrowest ring. Once the axes are drawn, plot the data in a single color. Connect the points with a smooth line.

f. Repeat c, d, and e for each group of students. Each group should measure their own spoke, and plot the data on a shared graph, each in a different color. Be sure the graph includes a key so that it is clear which color represents each group.

Materials:
one (or more) cross sectional slices of a tree trunk obtained from a local lumberyard

ruler(s) with millimeter tick marks

felt-tipped marker

graph paper

4 different colored pencils

Materials:
fresh leaves

plastic bag and damp
paper towel to keep
leaves fresh

magnifying lenses

white paper

tape

Examining Plant Stomata

Introduction to activity

Producing leaves with more or fewer stomata is one way plants adapt to their changing environment. List other adaptations plants and animals have made to their environment.

Procedure

Collect some leaves. Use a magnifying lens to examine leaves' undersides for stomata.

Discussion

Do some of the leaves appear to have more closely spaced stomata than others? What other factors would contribute to the number of stomata observed?

Teacher's Guide to the Activity

Examining Tree Rings

In this activity students work in groups to measure the distance between rings on sectional slices of tree trunks, and then interpret their findings. Divide the surface of the wood in four, and label each line of the spoke from 1 to 4, representing student groups. Each group counts the rings along its spoke, measures the distance between rings, and then graphs the data. Each group's data should be plotted on the same graph, but in a different color. Be sure the graph includes a key so that it is clear which color represents each group.

Hint

• The class may choose to measure only the first 10 or 20 rings if measuring all would be too time consuming.

Introduction Guide

a. In each case, the exposed grain will be different. Scientists know there are many different ways to display an artifact, and that each display yields different sets of measurements and interpretations.

Procedure Guide

f. Repeat c, d, and e for each group of students. Each group should measure their own spoke, and plot the data on a shared graph, each in a different color. Be sure the graph includes a key so that it is clear which color represents each group.

Examining Plant Stomata

In this activity students examine the stomata on fresh leaves.

Introduction Guide

Fish that live deep in the ocean where there is little or no light tend to have bigger eyes than fish that live in the well-lit surface ocean. Plants that grow where water is plentiful tend to have shallower roots than plants that live where it is very dry and therefore have to probe deeply into the ground for moisture. Trees that live in areas with harsh winters must avoid the buildup of heavy snow that could knock them over. Some trees lose their leaves so the snow can't build up on them. Other trees have needles that don't collect snow, instead of leaves. Animals that live in colder regions have developed strategies to conserve heat such as blubber and fur for insulation. Humans who evolved close to the equator, where sunshine is strong, have dark skin and eyes for protection against the Sun's rays. Humans whose ancestors lived in higher latitudes tend to have lighter skin and eye color because protection against comparatively weaker sunlight was not as crucial.

Time:
Examining Tree Rings takes 10 minutes to prepare and 50–60 minutes to conduct; Examining Plant Stomata takes 5 minutes to prepare and 15 minutes to conduct.

Discussion Guide

Other factors that would contribute to the number of stomata observed are the types of tree leaves, the age and size of the trees, and the overall health of the trees. Changes in atmospheric CO_2 probably do not account for any differences observed, because the leaves are contemporaneous. That is, they grew at approximately the same time, or at least in the same season. The leaves were also probably obtained from the same neighborhood. To attribute the differences in stomata density to changing CO_2 concentration, students would need to compare leaves that grew at very different times or in very different places. Ask students to suggest collection times and sites that would satisfy these conditions.

Designing Science Lessons

In addition to trying the lessons presented here, you might want to introduce activities that are more closely related to students' neighborhood or community.

The study of almost any topic related to bodies of water fascinates most people, especially students. Even very young students are excited by activities that explain our watery world. Given this, you might introduce students to the marine sciences. Where are local aquatic systems located? Is there a pond, lake, stream, river, or water reclamation system nearby? Do the sides of a town fountain show signs of plant or animal life?

Oceanography is one field in which students can wield the same tools practicing scientists do: close observation, predictions, data collection and organization, and communication. When presenting already-planned science lessons, or when designing new ones, consider having students learn the same way scientists do: by employing scientific inquiry.

Scientific Inquiry

1. Observing

In science, discovery begins with an observation. Scientists notice something of interest, something that seems curious, and ask, "Is this really what it seems to be? Why is it like this?" For this reason, learning to look closely is a fundamental skill for young scientists.

Be creative, even dramatic, in starting students off on inquiry methods. Present an intriguing object or event. Describe it in vivid terms, demonstrate it, or both, and help students record their observations. What do students notice? What strikes them? What is surprising and unexpected? What seems familiar? What is recognizable: color, shape, size, scent? Is anything predictable? Challenge students to identify sequences on a time line or in a cycle or, conversely, separate and distinct categories (tiny versus huge, dim versus bright, few versus countless).

2. Asking Questions

An investigation can take many possible pathways. What are students most curious about? What questions do they have? Pinpointing what students most wish to learn about a topic is very easy to do. The tools at your disposal include questions, a short vocabulary list, a reading, picture, slide or a videotape clip. You can use these to start a brainstorming, conceptual mapping, opinion charting, or dramatic session.

What do students want to find out? Have them show you via captioned drawings or illustrations done on paper or on the classroom floor (with masking tape), models (of anything from fish to rockets) made from paper, clay, or blocks, and so forth. Note the questions students put to a lively guest speaker who knows about your topic. (Photographers, artists, reporters, and other observers make wonderful guests. Students love to ask how photographs or videotapes were set up and shot, how animals were spotted and pursued, what local personages were interviewed for a story, and so forth.) What do students still want to know after reviewing material for inclusion in or exclusion from the classroom science library?

In short, to start off on an investigation, ask for students' questions and vary the ways in which those questions are presented.

3. Predicting Answers

What do students think they already know about your research questions? The work they generated in posing their questions will tell you something about this. Students' hypotheses, or predictions, about possible answers to their research questions will tell you more.

Explain to students that random predictions in answer to research questions are not very useful to scientists. In real experimentation, random guesses waste time and money, so scientists work on the possibilities they feel most confident about. Even young scientists can rank hypotheses as to the likelihood of their being accurate. They should also be able to explain why they ranked the hypotheses as they did. On what bases are students judging the credibility of predictions? Were their impressions formed by direct experiences? Have students encountered apparent proofs? If so, how often have they seen these? Is the seeming evidence indirect, via television, books, classroom lectures, parents, or friends?

With your students, arrive at a way of judging the credibility of hypotheses.

4. Designing Procedures

Scientists prepare for their experiments by planning them. They must decide what information to collect; choose and gather the equipment they will need; and prepare a means of recording data. Often, they test their instruments and practice their techniques before embarking on the experiment.

Deciding on what information to collect. Not every research question can be tested. Even professional scientists are limited in their access to materials and equipment needed to answer questions. Likewise, students should be helped to differentiate between the questions they're able to try to answer and those they aren't, given the limitations of their resources.

Choosing and gathering equipment. Designing a step-by-step plan to test predictions is a lot of fun. Students need only state exactly what needs to be done in the order it will be done. What materials are needed and in what quantity? Who will write down directions, check them, and set up any apparatus needed? (That is, who will be the research manager?) Who will take the role of investigator and carry out the directions? Who will record the results? Who will report them to the class?

One caution: It is so easy to leave information out of the research plan that even professional scientists rework their plans many times before beginning an experiment. Students will be amazed at how often their plans have to be revised before they are so clear that another person can duplicate conditions and procedures exactly, in every detail, simply by following directions.

Recording and organizing results. How will your students make note of what they see? Will they count and make tallies, measure, draw, or tick off items on a checklist? Will they make a graph? Will they draw their results and add captions to the drawings? Many students enjoy recording their observations orally, using an audiotape cassette player or video recorder. If drawn or written, students' findings might be presented on a variety of paper supplies, such as drawing pads, poster board, index cards, and lab books or journals. Students enjoy working with "unorthodox" paper and non-paper materials.

5. Executing Procedures

Many teachers rely on duplicate science stations to allow all the students in a classroom to conduct an investigation at the same time. Each station should be equipped with materials chosen and gathered; the agreed-on procedures; a list of words students might need in referring to the investigation; and the materials to be used in recording results.

It is often helpful to stand at one station and summarize or demonstrate an investigation for the class before the students themselves launch it. You might ask one student to read the directions aloud (and the rest of the students to read along silently) as you describe or carry out each step. When all students understand what they must do, let small groups conduct the investigations at their station. If you wish to divide the class so that some students work at the stations while others are occupied with other activities, set up a rotation schedule so all groups have a chance to conduct the investigation. Stress, however, that the research is being conducted by more than one small group for an important reason: procedures must be carried out several times before the results can be considered dependable.

6. Communicating Results

Data collected in the course of an investigation are generally presented in two forms: written and pictorial. Both are summaries of the work conducted. In written form, a standard format is used wherein students report their hypothesis, procedures, results, and questions (or suggestions for further research) that arose in the course of the research. Pictorially, a chart or graph presents the results in such a way that they can be taken in at a glance. Both offer excellent opportunities for students to reflect on what they've learned. They also afford you the chance to point out that science draws on many skills, including writing, drawing, reading, and mathematics, among others. Emphasize the broad applicability of this experience to other classes and to life overall, and encourage students to think likewise.

Easy Extension Activities

Exploring evidence for global climate change can carry you and your students into many fields of study related to science. These include geography, sociology, mathematics, language arts, and graphic arts. The following activities are easy starting points, and a comprehensive resource listing on pages 138–147 offers additional useful information.

ANIMAL BIOLOGY

1. Fossils are climate indicators because they come from organisms that have adapted to survive in an area. For example, monkey fossils suggest that warm-climate primates found a region to be habitable at one time. What supporting information would corroborate or disprove this deduction? Scientists would search for other evidence, such as the remains of trees and fruit. From these clues, they could deduce past vegetation levels, amounts of precipitation, surface temperatures, and levels of atmospheric gases. What can students deduce from hand, fist, or foot imprints in damp sand? Chart the features for which the imprints show some evidence, such as foot size, length of stride, body size, and the distribution of legs. Discuss features for which there is little or no evidence, including skin color, gender, diet, and protective features.

2. What *can't* we tell from fossils? Most living things decay and break up before they can become petrified. Of the small percentage that do become fossils, most are lost to the effects of time, weather, and marauders. Of the remaining few, most have not been and probably will never be found. Obtain a whole chicken or fish and draw it. Then remove (or cook off) its skin, muscles, ligaments, eyes, and feathers. (These would be too soft to survive the petrification process.) What information can be gathered from the remaining skeleton? What can be deduced from the evidence at hand? What is impossible to know? Create a plaster cast of some or all bones. Is more or less information available from imprints?

3. Unless ambient temperatures reach extreme highs or lows, human beings are able to maintain an internal temperature of 37°C. Many marine animals cannot maintain a constant body temperature because they have no means of insulating themselves against even small variations in water temperature. They overheat or become chilled unless they restrict themselves to certain localities. For them, body temperature is largely set by the temperature of the water in which they live. Marine mammals, whose heavy, built-in wetsuits of thick blubber protect them against extreme cold, can become overheated outside the temperature zones in which they live.

To see that body temperature is often a function of ambient temperature, prepare two bowls of water, one icy and the other hot (but not boiling). Use a thermometer to take the normal temperature of students' hands, then place one hand in each of the two bowls, respectively, for one minute. Take temperature of the hands after each immersion. How have they changed? If water temperatures warm too quickly for marine organisms to adapt, what difficulties might the animals encounter? Can students list other specialized adaptations to habitat conditions that might be threatened by changing climate?

4. Show animals' adaptations to food sources in their habitat, speculating on threats to survival posed by climate change. Use chopsticks, straws, spatulas, popsicle sticks, clothespins, scissors, spoons, tongs, fine-mesh strainers, and hammers to simulate the mouth configurations of various birds and fish. Use these tools to pick up pretzel sticks, fish crackers, flat crackers, popcorn, nuts, frozen peas, cocoa or instant milk powder. Which "mouths" are most successful at collecting which foods? How useful are such mouth parts if the animals' food sources migrate or die?

5. Detect and study aquatic organisms by using small, clean jars (or a plankton sieve) to obtain small samples of local fresh or seawater from a stream, pond, river, lake, ocean, or town fountain. Study these samples under a microscope. Ask students to devise a key useful in identifying the organisms they observe.

6. Many aquatic animals depend for survival on oxygen dissolved in water. Heating releases this gas, making less of it available for respiration. To illustrate this, seal transparent containers of aerated, boiled, and ordinary tap waters. When all samples have reached room temperature, create a small opening and add identical numbers of brine shrimp, water fleas, or goldfish to each container. Remove the animals as their movements become erratic. How do variations in dissolved-gas levels affect the animals' movements?

7. Animals and plants that live at the same time eventually leave their fossils in the same proximity. Fish, shell, and sea grass fossils found in the same location suggest that an area was once covered by water. Draw a five-layered core sample of sand, soil, and bits of stone in which organic matter (e.g., eggshell, other shells, bones, apple cores) are embedded. Try to match animals, plants and habitats convincingly (e.g., orange seeds, clam shells and pine needles don't go together). If there is sufficient time, the class can create sediment layers in clear glass or plastic containers.

8. Humans and their lifestyles often make areas unsuitable for many plants and animals to live. We remove the cover required for shelter, alter the water supply, change vegetation, and make noise that disturbs many animals. Can such threats be reversed? Can areas now unfriendly to plants and animals be made to afford co-existence with humans?

List animals that may have lived in the area your school now occupies. Clues can be found by studying nearby undisturbed areas, if any, or by interviewing members of local land-conservation groups. Look for evidence of smaller-sized animals (*e.g.*, birds, lizards, beetles, worms). Might any area of your school be suitable for conversion to microhabitats that these animals could use? For example, a corner or a nearby vacant lot might be made to entice wildlife. A clump of bushes, tall grasses, or native shrubs might be planted; rock piles arranged with space for animals to crawl and hide; and water bowls and seed piles placed under rocks. Check the area periodically and note how many types of animals are found in all possible microhabitats, including under plants, on leaves, in grass, in bark, and among rocks.

CHEMISTRY

1. To illustrate that the air around us contains invisible gases, request a demonstration of radon or carbon monoxide detection devices from the Occupational Safety and Health Administration (OSHA), fire departments, or other environmental safety programs.

2. Atoms contain positively and negatively charged particles that behave like the north and south poles of a magnet. Use the attractive and repulsive forces of magnets to model the forces at work at a molecular level in gases. Test a variety of materials (*e.g.*, iron filings and table salt) to determine whether or not they are magnetic, then magnetize as many as possible with bar or horseshoe magnets. Does a magnet's power to attract objects extend through other substances, such as air, water, oil, cardboard, thin glass, and metals?

3. All living things require nutrients to live and grow. Plants absorb theirs from soil or water, taking in nitrogen, phosphorus, potassium, calcium, sulfur, and magnesium, iron, chlorine, copper, manganese, zinc, molybdenum, and boron. Many nutrients come from the remains of dead plants that decay into the soil around them. When crops are grown for food, however, the plants are not returned to the soil. This disruption of the natural process of growth and decay depletes nutrients in the soil. The soil becomes useless for agriculture after a limited number of growing seasons.

One solution is to use chemical fertilizers to replace the nutrients artificially. Obtain five to ten containers of various fertilizers. Analyze their ingredients label and claims for performance. What standard features emerge from the comparison? (Fertilizers are usually labelled with a formula that indicates the percentage of several essential elements. A 10-8-5 fertilizer, for instance, is one containing 10 percent nitrogen, 8 percent phosphoric acid, and 5 percent potassium.)

4. Most living organisms depend on the chemistry of their habitats. Their well being is linked to the temperature-regulated rates of many chemical reactions. Even their body temperature is largely a function of environmental temperature.

In the event of planetary warming, chemical reactions such as those that make up the metabolism of plant cells, could speed up because rising temperatures excite atoms, creating instability or causing ruptures in the bonds holding molecules together. Most chemical reactions proceed more quickly at higher temperatures.

To examine these processes, pour equal amounts of very hot, tepid, and cold water through filters containing identical amounts of ground coffee. How does the water differ in color? How do students explain the differences?

5. Compare the density of saline and fresh water by coloring each with food dye before pouring them into a transparent container. Which water sinks to the bottom? Which rises to the top? What are the effects of stirring the mixture? Conduct a comparable experiment with very cold and warm water. What results do you obtain?

6. "Fix" fresh water contaminated by salt water by filtering, freezing, or distilling it. Filters, which are available in camping equipment stores, work by pushing water molecules against a screening membrane, leaving the larger salt molecules behind. In contrast, freezing, evaporating, and distilling take advantage of the fact that the freezing and evaporation processes leave behind salts and minerals.

Engineer a solar still by placing a small tray of salt water on a larger empty tray. Erect a transparent cover over both, and place them in a sunny spot. Soon, heat from the sun will evaporate the salt water from its plate. Water vapor will rise, condense on the inside of the cover, and run down into the larger collecting tray.

Discuss difficulties students envision in conducting these processes on a large scale (*i.e.*, to supply potable water for an entire city).

GEOLOGY

1. Four hundred years ago, Francis Bacon created a jigsaw map of the world showing that the continents of Africa, North America, and South America could be fit together. Since then, geologists have observed that many rocks and fossils on each side of the Atlantic Ocean are identical. This matching of rock layers is true in other parts of the world, so scientists now think the Earth's land mass was once combined into a huge continent called Pangaea (pan-GEE-uh). Dry some mud on a cookie sheet, then press down on it until it cracks. This is comparable to land masses having irregular shapes yet still fitting together.

Scientists have devised the unifying theory of plate tectonics to explain Pangaea's breakup. Geologists believe the continents ride atop six large plates and a number of smaller ones. The plates move slowly across the Earth's surface at rates ranging from 1–20 cm/yr, carrying the continents along with them. This movement has implications for climate change because land masses influence the

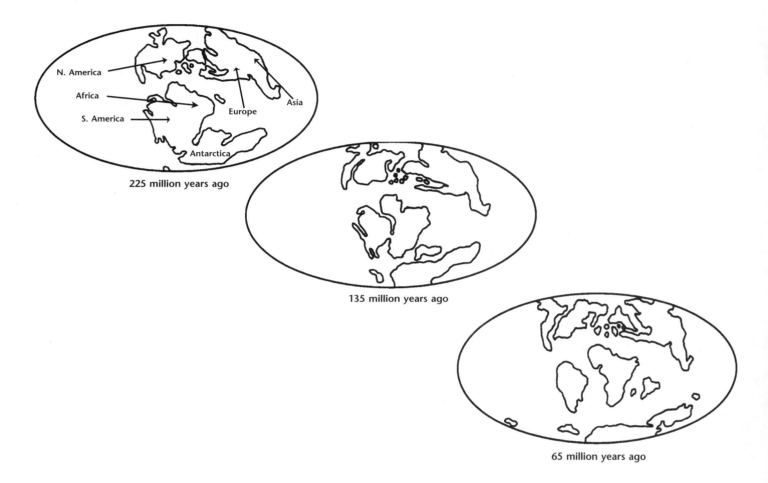

225 million years ago

135 million years ago

65 million years ago

ways in which winds and current distribute heat around the planet. A region's location on Earth affects the amount of the Sun's radiation it receives, absorbs, releases, and reflects.

Make a photocopy of or trace a world atlas. Cut out the continents and place them in their current locations around a beach ball. Use the guides given above to show where the continents were located and what they looked like during different periods in Earth's history. Move the cutouts accordingly. Try to predict the continents' position 50 million years from now. (Hint: Australia is moving north at a rate of almost 8 cm/year. Someday, it will lie wholly within tropical latitudes.)

2. When scientists find artifacts or imprints in soil, mud, ice, or rock, the materials are seldom in perfect condition. They have been so degraded by time, the elements, and animals that they are nothing more than a fragment of the original. Have students press their hands into a tub of dry sand. What kind of

impression do they leave? What happens if wet sand or clay is used? Subject the imprints to various destructive agents: "rain" or "floods" from a perforated paper cup, watering pot, or hose; wind from a paper fan, electric fan, or hair dryer; abrasion from sandpaper representing the rubbing action of shifting sands; digging from a spoon representing a dog, shovel, or bulldozer; and strong shaking to represent disturbance from earthquakes. In what condition are the imprints now? What inferences, mistaken or otherwise, might be prompted by materials such as these?

3. Geologists, paleontologists, biologists, and other scientists use equipment that drives a long cylinder into the ground. This cylinder is called a core, and the sand, soil, or mud it brings up is a core sample. Core samples may contain rocks, plant remains, and/or animal remains that have accumulated over the ages. Scientists analyze these materials for clues about the climate conditions over the millennia they were deposited.

Students can investigate these processes by driving a transparent straw, aquarium tubing, or a plastic syringe into an opaque cup filled with alternating layers of chocolate instant pudding, finely crushed vanilla wafers, graham crackers, chocolate sandwich cookies, raisins, and two sizes of chocolate chips. Assuming one millimeter equals sediment deposition over 10,000 years, what information might various layers convey?

4. Using various types of uncommon, interestingly-shaped pasta (*e.g.*, farfalle, radiatore, and cartwheels with spokes), engage in the analysis and deductive reasoning scientists must pursue when they find only a fragment of a specimen or artifact. Begin by presenting students with very small pieces of pasta and ask them to deduce as much as they can about each object. Repeat this exercise by presenting the students with ever-larger pieces of pasta. Encourage students to fit the pieces together as if they were assembling a jigsaw puzzle. If given more than one type of pasta simultaneously, can students tell them apart?

5. Initiate a rock collection, performing chemical and other tests to identify age, composition, rate of deposition, and other characteristics. A field guide can be used to assist with this activity.

6. Coastlines form as a result of a combination of geologic, geographic, and meteorologic factors. Discuss the different types of coastlines that exist, and how they change naturally over time. List some of the ways humans have used the coastline for housing, recreation, and business activities. What impacts could these activities have on the natural evolution of these areas? Are these positive or negative effects? What can be done to protect low-lying areas against rising seas? Are these temporary or long-term solutions?

7. Collect soil samples from a variety of locations and seal them in glass containers. Place the containers in a sunny spot and leave them undisturbed for

several hours. In which jars does the most water condense? Which soil contained the most water? What features of the soils seem to correlate with their water retention: color, particle size, presence of organic matter, and location collected?

METEOROLOGY

1. Discuss what weather events most influence living conditions, recreational choices, and agriculture in your region. How do students rank the importance of wind, fog, drought, heavy rains, seasonal rains, permafrost, and the like?

2. Use a camera's light meter to demonstrate differences in the reflectivity of various ground surfaces or coverings. Assemble comparably-sized containers of dark soil, wood and bark chips, green leaves and grass, light-colored sand, water, and ice. Place these in a location such that each receives the same amount of strong Sun. How much light is reflected from each? Notice that the camera's shutter speed or F-stop must change from a low F8 for dark surfaces to a high F22 for highly reflective ones to accommodate the varied amounts of light.

3. Why are clouds different? Why do some look puffy, some wavy, and some straight? Some clouds are made up of liquid water droplets, while others are made of ice crystals. Clouds made of water droplets tend to have edges that look fuzzy, while clouds made of ice crystals usually have sharper edges. In addition, the amount of water or ice in clouds varies. Thick, opaque clouds hold more water than thin, wispy ones.

Clouds also form at different altitudes. That can make them appear different to a person on the ground. But perhaps the most important reason clouds look different is that the physical processes by which they form, and the forces that move them after they form, are different. Vertical motions—up drafts and down drafts—produce puffy cumulus clouds. Where vertical motion is weak or absent, layered stratus clouds are more typical. Horizontal motions can also smear out clouds, making them thin and wispy (cirrus clouds).

Document differences in clouds' number, color, movement, size, and shape over a short or longer period of time. Can students recognize the primary cloud types of cirrus, cumulus, stratus, cumulonimbus, and nimbostratus? Consider using unorthodox materials such as cotton balls left intact or pulled apart into wisps.

4. Observe talcum powder sprinkled on an illuminated light bulb or the ashes from burning paper to see that heated air rises into the atmosphere, carrying particles with it. Alternately, put a small empty ice chest in the freezer overnight. When it's time to open the chest, group students together and pour the cooled air over their hands. Dense and heavy, cold air will sink through warmer air. This is the principle on which open, trough-shaped supermarket freezers work.

5. Create a simple thermometer to monitor rising and falling temperatures. Fill a straight-sided, glass jar (*e.g.*, an olive jar) half-full of colored water. Cover the jar

with aluminum foil secured with a rubber band, and drive a drinking straw through the center of the foil so that the straw is suspended in the center of the jar. Move the thermometer toward and away from different sunny and shady spots or leave it in one place that will experience changing conditions. (As an alternative, put the thermometer in a protective box out of direct sunlight—sun, wind, and rain affect true temperature readings.) Watch the water level in the straw rise and fall accordingly. This thermometer can't record degrees, so take a real thermometer along to corroborate and refine your observations.

6. How much rainfall is your neighborhood getting? Create a rain gauge by placing a funnel into a transparent, straight-sided glass jar. Put the jar in an area clear of bushes, trees, and overhanging roofs, and monitor precipitation levels.

7. To make a wind vane, you will need two index cards, scissors, a stapler, an unsharpened pencil, a push pin, and a plastic straw. Cut a triangle out of one index card and fold the other in half. Staple the triangle to one end of the straw and the folded index card to the other end. Use the push pin to make a large enough hole in the straw's balance point that the straw will be able to rotate freely, and pin it into the pencil eraser.

Go outside, and hold up the wind vane. In which direction does the arrow point? Use the wind vane repeatedly, at the same time each day or at different times, to learn what wind directions prevail in your locale.

8. To construct an anemometer, you will need four index cards, a red magic marker, two plastic straws, a stapler, tape, a push pin, and a new pencil. Color one card red, then cut four 3 x 3 cm crosses out of all the index cards. Staple the centers of the cut cards to the ends of two straws (two per straw), then fold and tape the index cards into small cups. Find the balance point of both straws, attach them at these points by stapling them together, and use the push pin to pin the straws into the pencil eraser. To measure wind speed, count the number of complete rotations per minute of the red cup and record the results.

PHYSICS

1. Compare the temperatures within and immediately outside a "greenhouse" (*i.e.*, a soda-bottle terrarium) by placing thermometers in both locations. Shield both thermometers from direct sunlight. Alternately, call local greenhouses and inquire as to their internal temperature, particularly on very cold days. Similarly, compare the heat conserved by various materials (*e.g.*, water, sand, soil, grass, rocks) when covered and uncovered.

2. Varied localized and global mean temperatures of Earth's atmosphere result from the sort of interactions that computer spreadsheets are designed to capture and display. Learn how to use the data tables and graphs in spreadsheet programs to conduct simple climate modeling exercises. Although spreadsheets—

and, therefore, students' models—are limited in assuming essentially limited relationships and are realistic only within a rather small range of temperatures, even professional climatologists and physicists struggle with comparable limits in their own models. Once familiar with spreadsheet functioning, refine the models by adding levels of interactivity (*i.e.*, feedback).

For data sets to use with your spreadsheets, refer to the World Wide Web sites listed in the Annotated Bibliography, or other federal or university Web sites.

3. Create a solar-energy oven by removing the top of any small-to-medium sized box and lining the box with aluminum foil. To fashion a spit suitable for suspending and cooking food, straighten out a coat hanger and poke it through two parallel sides of the box. Place a thermometer inside the oven to measure temperatures at various times of the day or under various conditions.

4. To demonstrate that heated air rises into the atmosphere, fit a balloon over the mouth of an empty plastic bottle. Place the bottle into a bath of hot water. When the air inside the bottle has been warmed, it will rise into and fill the balloon.

5. Compare the amounts of heat absorbed by surfaces of different colors by conducting an informal survey in the school's parking lot. How hot are the hoods of differently-colored cars relative to each other?

6. Earth's atmosphere surrounds the planet and penetrates into the ground. It consists of the mixture of gases we think of as "air," and has been divided by scientists into four layers. The troposphere rises 18 kilometers from the ground, and is the layer in which all life and almost all weather take place. The stratosphere extends upward another 22 km, to 50 km. Approximately 99 percent of the air in the atmosphere is found in these first two layers.

The mesosphere is far enough away from Earth that atmospheric temperatures fall to their lowest point: about -90°C. There, the sky is black instead of blue. The highest level, the thermosphere, begins 90 km off the ground. Temperatures rise, sometimes to 800°C, due to gases' absorption of short wave ultraviolet radiation.

Simulate reception of the Sun's energy on Earth through the atmospheric layers by wrapping layers of plastic wrap around a globe in thickness corresponding to the height of each layer. Use the visible light of a flashlight to model the full spectrum of solar radiation. Can students calculate placement of the flashlight to correspond to the Sun's actual distance from Earth (if Earth is a standard-sized classroom globe)? Describe the amount of light that reflects from the plastic wrap as well as that which reaches various areas of the globe. For greater verisimilitude, create varied cloud cover by inserting small pieces of tissue paper and paper towel between layers of the lower atmosphere.

7. Doppler radars send radio waves into the atmosphere in wavelengths of approximately 10 cm (which approximates the wavelength used to cook foods in microwave ovens). When these radio waves encounter obstacles such as droplets of water, hail, snow, tornadoes, birds, insect swarms, or airplanes, they change frequency. The radar's antenna scans for all returned signals, and experienced technicians interpret the differences in signals' strength or intensity.

Police traffic-speed radars also work this way. Invite a local Highway Patrol officer to display his or her radar device. If possible, follow the officer to a well-trafficked street bounding your school to see a demonstration with moving cars. Record all observations.

8. Consider the Goldilocks Problem: Venus and Mars, our nearest neighbors, are respectively far too hot or too cold to support life. Only Earth, to our knowledge, is just right. Why is this so?

Most life on Earth either resides in liquid water or is largely made up of it. Students know that water exists in solid, liquid, or gaseous states depending on its temperature (and the pressure exerted on it). The average global temperature is approximately 15°C on Earth, 480°C on Venus, and -60°C on Mars. Given this, is liquid water likely to be present on the latter planets? (No, though there is frozen water, frost, and water chemically combined with other substances on Mars.)

The amount of energy radiated by the Sun is not constant. For instance, the Sun was 25–30 percent fainter several billion years ago than now, yet Earth was warm enough then to support liquid water and the origins of life. Can students explain this apparent contradiction? (There were larger concentrations of greenhouse gases in the atmosphere at the time. Similarly, on Venus, thick clouds of sulfuric acid reflect away much of the solar energy the planet receives from the Sun. Venus receives only 90 percent as energy as Earth at its surface, yet is far hotter than Earth because of its own very strong greenhouse effect.)

PLANT BIOLOGY

1. Soil is a complex mixture of inorganic materials weathered from rocks, dead organic matter in various stages of decay, and living organisms. In fact, one single teaspoon of soil may contain five million bacteria, 20 million small fungi, and one million protozoa! These living organisms break down complex organic material into simpler molecules that can be used by plants.

Identify plant matter, such as leaves, stems, seeds, and bark in soil samples removed from various locations. Can students devise ways of calculating and explaining the relative proportion of organic to inorganic matter in each sample?

2. What means can students devise to document plant growth over time or under various conditions? Outdoors, show seasonal changes in plant stands—

including lawns, crops, river reeds, marsh grasses, pond algae, woods—via photo or drawn observations. Indoors, monitor potted plants or a sealed terrarium for evidence of plant respiration or transpiration. Conduct simple experiments: plant several groupings of herbs or vegetables in separate but identical soda-bottle terrariums, then vary air and/or water temperatures, or test how salt water solutions of various strengths affect plants' growth. Do certain species of plants tolerate saline solutions better than others?

Students may notice that some plants grow faster than others. These differences may be explained by the availability of nutrients, differences in water retention, and so forth. However, less obvious factors such as genetic traits may also cause differences in plant growth.

3. If the classroom has no potted plants, bring several from home for students to water. Notice that some liquid drains out the bottom of the pots. This liquid is not pure water, even in the case of plants given only water to which no fertilizer was added. What emerges from the bottom of the pot is a solution containing substances leached from the soil and, in the case of fertilized plants, excess fertilizer. Fertilizers applied to commercially-grown plants also end up in the runoff water. If these nutrients cause blooms of algae in nearby waterways, the ecological balance of these areas may be disturbed.

Collect runoff from classroom plants watered with and without fertilizer. Pour identical amounts into glass jars, placing these in a sunny spot. After several days, the runoff that contains fertilizer should begin to show growth of algae. This is particularly true of plants fertilized organically (*i.e.*, with animal products, which contain high levels of living microorganisms). Over a longer period of time, the differences should be quite distinct. Challenge students to devise a way of characterizing them.

4. Maintain a classroom, cafeteria, or school compost pile in a container at least 10 x 10 cm high. From what materials could compost be made? Any organic, decomposable matter can be used except for animal products such as chicken skin or bones and materials rendered toxic by herbicides or pesticides. Combine equal parts of "brown and dry" (*e.g.*, dry weeds, leaves, sawdust, nut shells, small sticks and twigs, shredded unprinted paper and cardboard, ashes) and "green and wet" (*e.g.*, grass and shrubbery clippings, fruit peels and cores, eggshells, seaweed). Keep the pile as moist as a wrung-out sponge, cover it with a tarp, and mix it about once a week.

Where could the compost be placed to fertilize plants at your school?

5. Think about all the trash we make in one day. What are the contributors to students' personal trash pile? Many of the items are probably derived from wood and wood pulp (*e.g.*, boxes and wrappers from lunch, used paper, bus tickets, tissues, old calendars and phone books, magazines, paper towels, napkins,

wrapping paper, grocery bags, and discarded juice containers). Consider creating usable recycled paper such as holiday stationery, cards, or school business cards from these discards. Blend 1.5 liters of water and a handful of shredded paper in batches in a blender, repeating once or twice to have enough pulp for a 30 x 30 cm sheet of paper. Add decorative elements such as grass, leaves, food coloring, or glitter, then spread the pulp on a piece of screening and let dry.

Earth Through Time

Since an average person lives fewer than 100 years, it is difficult to think about periods of time as long as 5,000,000,000 years. Let the age of the Earth, 4.6 billion years, equal the diameter of the Earth, about 13 million meters. Then the age when dinosaurs thrived about 100 to 200 million years ago corresponds to approximately 425 km, about the distance from San Francisco to Santa Barbara. The coldest part of the last ice age, about 20 thousand years ago, corresponds to a distance of about 60 meters, shorter than the length of a football field and roughly the height of a fifteen-story building. The 100 years since the publication of the first theoretical paper on the greenhouse effect can be represented by a length of about 30 cm, a little longer than an average sheet of notebook paper. One year, during which 100 million more humans will live on Earth, can be envisioned as about 3 mm, about the width of the E in the word Earth.

On the following six pages are a schematic of Earth history through the 4.5 billion years of its existence. Six bars represent the timeline. The grey areas on each bar end represent the amount of time covered by the following bar.

Bar A: 5 billion years through 5 hundred million years ago

Bar B: 5 hundred million years ago through 5 million years ago

Bar C: 5 million years ago through 50 thousand years ago

Bar D: 50 thousand years ago through 500 years ago

Bar E: 500 years ago through 50 years ago

Bar F: 50 years ago through 1 year ago

✳ Of the 51 events mentioned on the timeline, 32 of them occurred over the last 100 million years—within only the most recent 2 percent of Earth history.

✳ Humans have existed for only .05 percent of Earth history.

✳ Of the 51 events mentioned on the timeline, 21 of them occurred in the last 10 thousand years—within only the most recent .0002 percent of Earth history.

Ⓐ Years Ago x 1,000,000,000

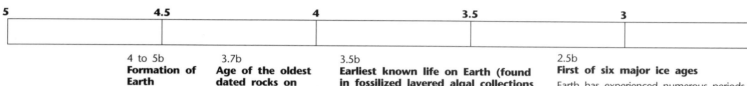

| 5 | 4.5 | 4 | 3.5 | 3 |

4 to 5b
Formation of Earth

3.7b
Age of the oldest dated rocks on Earth (found in Greenland)

Various radioactive isotopic techniques are used to determine the age of a rock. Fossils preserved in rock layers also are used to estimate the time the rock formed.

3.5b
Earliest known life on Earth (found in fossilized layered algal collections called stromatolites in Western Australia)

No one knows precisely how life originated, although several theories exist on how the first organic molecules formed. Life probably began in the ocean, where fundamental elements combined to produce carbon-based molecules.

2.5b
First of six major ice ages

Earth has experienced numerous periods of glaciation which have alternated with relatively warmer (interglacial) intervals. Periodic changes in the tilt of the Earth's rotational axis, in Earth's orbit around the Sun, and in the intensity of solar activity may have triggered ice ages. Changes in atmospheric composition, especially CO_2 concentration, also may have enhanced climate change.

Ⓑ Years Ago x 100,000,000

| 5 | 4.5 | 4 | 3.5 | 3 |

500m
Appearance of jellyfish-like animals and primitive fish

450m
Fourth of six major ice ages

440-405m
Earliest land plants and amphibians

Land plants further enhance the removal of atmospheric CO_2 through photosynthesis. The physical and chemical breakdown of rock by plant roots begins to form topsoil and allows greater areas of land to be colonized by plants. Land plants also play an important role in the hydrologic cycle by taking up and releasing fresh water. In certain conditions, plants which died and were buried under layers of sediment and rock were transformed into fossil fuels such as coal and oil.

320m
First forests appear

Forests take up and store carbon, generate and retain nutrient-rich soil, provide habitats for numerous organisms in addition to trees, and affect regional climate.

300m
Development of polar glaciation, fifth of six major ice ages

The retention of water as ice at the poles causes sea levels to drop, exposing once-submerged continental shelves, altering the paths of ocean currents, and reducing the water vapor content of the atmosphere.

350-250m
Supercontinent Pangaea coalesces

Plate tectonic rearrangement of continental land masses may have caused changes in the strength and direction of ocean currents. Changes in the way currents distribute heat may have caused major climate change, which in turn could have effected the distribution and evolution of plants and animals.

2	1.5	1	.5

2.1-1.2b
Marine algae and marine worm burrows are found in rocks of this age

Photosynthetic activity by marine plants took up CO_2 from the Earth's early atmosphere, which was rich in this gas. Eventually, enough O_2 was released to the atmosphere to allow respiration and the proliferation of other life forms.

1b
Second of six major ice ages

700m
Third of six major ice ages

2b
Concentration of atmospheric oxygen begins to increase

Aerobic organisms begin to evolve.

2	1.5	1	.5

250m
Primitive reptiles appear; majority of plants became extinct; majority of marine animal species become extinct

Several major extinctions are recorded in that fossils of many species gradually stop appearing in rock layers. Causes of mass extinctions are unknown, but their timing and duration are generally understood. This particular extinction was not sudden; instead, it was spread over millions of years. It is thought that the formation of Pangaea reduced the area of shallow-sea habitats, thus contributing to the extinction.

225-180m
Rise of dinosaurs; appearance of primitive mammals

180m
Breakup of Pangaea begins

The rearrangement in land masses that were once connected isolates species in detached regions. The isolated populations evolve independently, as occurred in Australia.

160m
Toothed birds appear; rise of flowering plants

This period was characterized by the evolution of increasingly complex species.

75m
Average global atmospheric temperature was 5.5°C warmer than present

The global temperature has risen and fallen repeatedly over geologic time in response to natural, pre-human forces.

75m
Environmental disruption— evidence of widespread fires, followed by interval of high precipitation

The fires might have been the consequence of high atmospheric oxygen concentrations. This represents a major change from the oxygen-poor atmosphere of the newly-formed Earth. Evidence for fires is preserved in wide-spread, ash-rich rock layers, which can be dated. Evidence of subsequent flooding is also found in the rock record, as layers showing evidence of erosion and sediment transport by water. Another possible cause of fires is increased volcanic activity, which is also linked to atmospheric changes.

65m
Mass extinction of dinosaurs and other forms of life

This extinction might have resulted from a climate change precipitated by a meteoric impact. The impact could have produced enough atmospheric dust to block the Sun's warmth.

2m
Whales, apes, and grazing animals appear

C Years Ago x 1,000,000

5	4.5	4	3.5	3

2.8m
**North and South
America joined by
the closing of the
Isthmus of
Panama**

Changes occur in
world ocean
circulation pattern
and heat transport.

D Years Ago x 10,000

5	4.5	4	3.5	3

	2		1.5		1		.5				

2.5m
Early humanoids appear

1.5m
Last of six major ice ages begins

.7m
Most recent change in polarity of the Earth's magnetic field

The orientation of the Earth's magnetic field has reversed numerous times, although it is unclear why this happens. Mineral grains in sediments and molten rock align themselves within the existing field. By analyzing the position of the magnetic minerals in old rocks, and dating those rocks, geologists can reconstruct the timing of the magnetic field changes, and the history of how some crustal plates have moved over time. Mirror-images consisting of rows of alternating magnetic stripes on both sides of a mid-ocean ridge were some of the earliest evidence of seafloor spreading.

130-120t
Global temperatures have fluctuated

Evidence for fluctuations comes from pollen in ice cores.

	2		1.5		1		.5				

20t
Sea level was 100 m lower than today

Oceans play an important role in regulating atmospheric composition; for example, they soak up excess carbon dioxide. A large decrease in ocean surface area and volume can alter the concentrations of atmospheric gases. A buildup of atmospheric CO_2 may cause warming, which could cause polar ice to melt, and sea level to rise. This is one example of the negative feedback which helps to stabilize climate.

18t
Coldest part of the last ice age: 20 times more dust/ash in the atmosphere than today

Atmospheric particles can block incoming solar radiation from penetrating to the Earth's surface. Particles can be natural (*e.g.*, volcanic) or man-made (*e.g.*, pollutants).

12t
Woolly mammoths, mastodons, saber-toothed cats, giant ground sloths, cheetahs, lions, zebras, yaks, tapirs, capybaras and spectacled bears thrived in North America

Fossilized bones of these animals, which do not currently live in North America, suggest that conditions at the time were much different than today.

10t
Last ice age ended

Though it was once thought that the onset and waning of ice ages occurred very slowly, newer evidence from glacial ice cores suggests that climate changes might have taken place relatively quickly—perhaps within decades. Such relatively sudden changes might not allow some species to adapt to new conditions, or even migrate to new habitats.

5t
Fossils suggest the area that is now the Sahara Desert was lushly vegetated and supported abundant animal life

How this area changed so dramatically is not well understood. Changing atmospheric patterns, such as wind direction and rainfall distribution, in concert with movement of the African plate closer to the equator, may have caused the transformation. Today, there is concern that the Sahara is expanding, that the land bordering the desert that is now used for agriculture is becoming more arid. But these recent changes might be due to unwise farming practices, which increase soil erosion and the ability of the ground to hold water and support vegetation.

1t to 700
Global warming trend (anthropological evidence)

Evidence for this warming trend comes from studies of human migration patterns. Areas of northern oceans that are now permanently frozen thawed temporarily, allowing sea voyages, then froze again, preventing further contact between separated populations. During this period, the Vikings colonized Greenland and then abandoned their settlement after the climate grew colder.

Ⓔ Years Ago x 100

5	4.5	4	3.5	3

545-145
Little Ice Age; period of decreased solar activity

This period, recorded in human history and characterized by severe winters and poor growing conditions in Europe, coincided with a time of decreased sunspot activity, which is an indicator of periodic decreased energy output from the Sun.

345-145
Doubling of world human population (from 0.5 to 1 billion)

Though a doubling in population over 200 years sounds impressive, the rate at which the human population is growing exponential, which means that the rate itself, not simply the number of people, is increasing over time. Humans, and their needs for food, water, energy and space, have the potential to exceed the planet's capacity to sustain their needs. Enhanced environmental degradation will almost inevitably accompany human population growth.

200
Discovery of photosynthesis

Photosynthesis, which removes atmospheric CO_2 and releases O_2 is an important proc in regulating atmospheric carbon dioxide concentration.

Ⓕ Years Ago x 1

50	45	40	35	30

45
Development of the radiocarbon (^{14}C) dating technique; discovery of the Mid-Atlantic Ridge

The ^{14}C method has allowed precise dating of biogenic materials that enable scientists to reconstruct climate change over thousands of years. Seafloor spreading centers are the locations where new ocean crust is produced, an important process in plate tectonics. Seafloor spreading along the Mid-Atlantic Ridge proceeds at about 10 cm per year.

37
First accurate atmospheric CO_2 measurements

Charles Keeling developed a method of measuring atmospheric CO_2 in 1958. Measurements since that time have shown an unmistakable rise in CO_2 due to human causes.

37-7
10 percent rise in atmospheric CO_2

CO_2 concentrations are now rising at a little less than 0.5% per year.

30
Beginning of the development of plate tectonic theory

2	1.5	1	.5

194
Discovery of infrared radiation

The infrared portion of the electromagnetic spectrum is felt as heat. Most of the energy radiated by the Earth is in the infrared band. This energy, trapped and re-radiated by atmospheric gases, warms the Earth.

200-present
Atmospheric methane concentrations increased from 650 to 1800 parts per billion by volume

Methane, a heat-trapping gas, has multiple sources, many related to agriculture (*e.g.*, rice paddies and the digestive systems of cattle). The increase in atmospheric methane, which now occurs at the rate of one percent per year, reflects the growing human population.

150-present
Atmospheric CO_2 concentration rose 25 percent

This increase resulted from the burning of fossil fuels, which intensified following the Industrial Revolution. About half of the CO_2 produced in combustion of these fuels stays in the atmosphere. The other half is absorbed by the oceans or biosphere.

100
Average global temperature was about 0.5°C cooler than present

Warming over the past century is consistent with the theory of an enhanced greenhouse effect, but by itself this observation is not evidence of a long-term global trend. Temperature measurements have not always been accurate. It is also important to note that temperature rise is not steady over time. There have been intervals of cooling, even though greenhouse gas levels have continued to rise over the same period.

100
First theoretical paper on the greenhouse effect published

Concern about enhanced global warming did not immediately follow.

83
Theory of continental drift (predecessor to plate tectonics) first published

Few geologists found this idea convincing, primarily because it provided no explanation for how the continents moved. It would be decades before modern plate tectonic theory was formulated and accepted.

65
Invention of chlorofluorocarbons

These chemicals are non-toxic and have many industrial uses, but they destroy stratospheric ozone (which filters harmful radiation) and are very effective heat trapping greenhouse gases. They have a long (100-year) residence time in the atmosphere.

20	15	10	5	4	3	2	1

20
Theory that CFCs destroy ozone published; oxygen isotope technique to determine paleotemperatures developed

Another decade passed before the depletion of Antarctic ozone was documented. Ozone is the only component of the atmosphere that absorbs damaging ultraviolet (UV-B) radiation. The isotopic signatures of water, ice, trapped gas bubbles and shell material in sediments are useful indicators of past climate change.

18
Discovery of seafloor hydrothermal vents

These vents occur along seafloor spreading ridges. Cold ocean water cools newly-formed crust. The heated water reacts with the rock, and this mineral-rich fluid spews from the vents. Animals which live around the vents use chemicals in the fluid to produce energy. At times in Earth's history, venting may have been more widespread and vigorous than today. The process, to varying degrees, influences the composition of seawater and the ecology of the deep sea.

5
Nations agree to phase out CFCs by 2000

Developing CFC substitutes and refitting factories will cost hundreds of millions of dollars. Because the decay rate of CFCs in the atmosphere takes many decades, even though CFC production has declined since 1987, the total amount in the atmosphere is large and still growing as existing CFCs are released.

5
Earth's human population reaches 5.3 billion (and is growing by 100 million each year)

Exponential population growth will further stress the environment by amplifying the enhanced greenhouse effect, loss of biodiversity, and ozone depletion.

1
There were roughly 5000 more species of plants and animals on Earth than there are today.

This rate of change—5000 fewer species a year—is about 10,000 times the natural loss rate! Production of food and medicine, as well as the ecology of many areas, depends on biodiversity. It is estimated that Earth is home to some 30 million species, of which humans have catalogued only 1.4 million. We are losing potentially valuable resources that we haven't yet discovered.

Altitude: distance above the Earth's surface, usually measured from sea level

Annuli: yearly concentric growth rings on fish scales

Anterior: area of an animal nearer to the head

Anthropogenic: produced or caused by humans

Aromatic: having an odor

Atmosphere: layers of gases that surround a planet

Atom: a primary building block of a chemical element

Biodiversity: the variety of plant and animal species in an ecosystem; the variety of Earth's ecosystems

Bond (in chemistry): force that holds two or more atoms together to form molecules

Botanist: scientist who studies plants

Calcite: the rock-forming mineral, calcium carbonate ($CaCO_3$)

Carbohydrates: compounds of carbon, oxygen, and hydrogen including sugars, starches, and cellulose produced by plants during photosynthesis; used during respiration to produce energy

Carbon dioxide: a naturally occurring atmospheric gas that contributes to the greenhouse effect; a molecule of carbon dioxide is composed of one atom of carbon and two atoms of oxygen (CO_2)

Chlorofluorocarbons (CFC): synthetic compounds that contain the elements carbon, chlorine, and fluorine

Chronological: system of organizing events according to when they occurred in time

Circuli: concentric ridges that compose the bony layer of fish scales

Climate: average long-term weather conditions in any particular region

Climate zones: divisions of the Earth according to similarities in climate

Climatologist: scientist who studies climate

Composition: the parts, or arrangement of parts, within a substance

Compound: a substance that is made up of two or more parts; a molecule made up of at least two elements

Concentration (related to greenhouse gases): amount of a gas within a unit volume

Concentric: having circles embedded within each other, as in the rings of a tree

Condensation: process in which a gas changes phase and becomes a liquid or solid

Configuration (related to greenhouse gases): the spatial arrangement of atoms in a gas molecule

Constancy: uniformity; the extent to which something remains unchanged

Contemporaneous: living or occurring during the same period of time

Contour plowing: farming technique in which agricultural land is plowed in patterns that minimize erosion

Coral polyp: individual marine animal living within a calcium carbonate skeleton; lives in colonies

Coralline algae: type of red algae that can form reefs

Core: long, plug-like cylinder of sediment or rock; extracted by scientists to understand geologic processes

Decompose: process in which a compound is broken down into simpler parts (*i.e.*, into its constituent elements by bacteria and chemical processes)

Deforestation: clearing of trees and forests

Deposition: the laying down of sediments (*i.e.*, the settling of rock particles and organic debris through water to a lake or ocean bed)

Diffuse: to disperse or spread out

Disperse: to scatter

Dissipate: to scatter

Earth's axis: an imaginary line drawn through the North and South Poles about which the Earth rotates

Electromagnetic radiation: energy within a spectrum of wavelengths emitted by the Sun, the Earth, and other planetary bodies

Element (chemistry): substance that cannot be separated into simpler substances by chemical means

Ellipse: a planar figure resembling a flattened circle

Emission: substance released or discharged (*i.e.*, exhaust from a car engine)

Equilibrium: balance of forces; the condition that exists when a chemical reaction and its reverse reaction proceed at equal rates

Equinox: time when the Sun crosses the plane of the Earth's equator and day and night are of equal durations

Eradicate: to completely destroy

Erosion: physical movement or chemical dissolution and transportation of soil and rock which can change the landscape

Estuary: zone where a river enters the ocean

Evaporation: process in which a liquid turns to gas

Feedback: processes that can amplify (positive feedback) or inhibit (negative feedback) progress of an initial action

Focus (with respect to fish scales): the point on a fish scale around which circuli and radii are arranged

Foraminifera (also **Forams**): a group of marine animals that produces chambered shells of calcium carbonate; these shells are deposited as seafloor sediments

Fossil fuels: deposits of organic material that have been converted to combustible fuels (e.g., coal, oil, and gas) under pressure by alteration and decomposition of plant and animal remains

Greenhouse effect: process in which gases in the Earth's atmosphere trap a portion of the Sun's energy and keep the temperature of the planet warmer than it would otherwise be

Greenhouse gases: gases that contribute to the greenhouse effect (i.e., carbon dioxide, methane, nitrous oxide, chlorofluoro-carbons, and water vapor)

Groundwater: water contained in the open spaces within rocks

Hemisphere: any half of a sphere or the Earth (i.e., Northern, Southern, Eastern, or Western)

Hydrologic cycle: the continuous flow of water in various states through the terrestrial and atmospheric environments (i.e., water evaporates from the Earth's surface, condenses to form clouds and precipitation, runs off the land surface to the ocean, and evaporates once again)

Hypothetical: involving an assumption or tentative theory that is subject to investigation and testing

Indigenous: originating in and characteristic of a particular region

Inert (related to gases): having little or no ability to react chemically

Inference: the process of drawing a conclusion by reasoning from evidence

Infrared radiation: electromagnetic energy in a specific wavelength range (longer wavelengths than visible light) that can be felt as heat; more than half of the incoming solar energy is within the infrared band

Infuse: to cause to penetrate

Insolation: the intensity of solar radiation on a unit area at a specified time on a particular surface

Isotope: a form of a chemical element with a different atomic weight than another form of the same element; weight differences reflect a differing number of neutrons within the atom

Light spectrum: the visible wavelengths of electromagnetic radiation, which humans perceive as colors of the rainbow

Mastodon: an extinct large mammal, resembling an elephant

Meteorologist: scientist who studies the atmosphere, weather, and climate

Methane: one of the greenhouse gases, consisting of one carbon atom

and four hydrogen atoms (CH_4); released to the atmosphere both naturally and anthropogenically

Microclimate: climate of a specific location, for example an "urban microclimate" affected by pavements, air pollution, etc.

Migrate: to move from one region or habitat to another

Molecule: a basic unit of matter composed of one or more elements; made of atoms held together by bonds

Nitrous oxide: one of the greenhouse gases consisting of two atoms of nitrogen and one atom of oxygen (N_2O); used as an anesthetic

Orbit: the curved, usually elliptical, path made by an object revolving around a celestial body

Otolith: a bone-like object in the ear

Particulate matter: finely-divided solid substances that can be suspended in or settle through water

Phase change (also: **Change of state**): process in which solids, liquids, and gases change from one form to another (i.e., ice melting to water, water turning to vapor)

Photosynthesis: process in which chlorophyll-containing plants use carbon dioxide and water, in the presence of light, to produce carbohydrates; oxygen is released in this process

Planetary tilt: the angle at which the Earth's rotational axis is tilted from vertical

Plate tectonics: the theory that the Earth's thin outer shell is composed of several discrete, rigid units (plates) that move separately with respect to one another

Posterior: the hind (rear) end of an animal

Potability: degree to which water is fit to drink

Precipitate: a solid that forms as a product of a chemical reaction in a solution, or that remains when water evaporates from a solution

Precipitation: any form of water (liquid or solid) that falls from the atmosphere and reaches Earth's surface (*e.g.*, rain, sleet, hail, snow)

Procedural error: inaccuracy in the methods used to gather or analyze data

Quadrat: a sample area enclosed within a frame

Quantity (related to greenhouse gases): an exact amount with specified units

Radiate: to emit, in rays, from a center

Radii: lines extending from the center of a circle outward to the perimeter; lines extending from the focus of a fish scale outward

Rampart: a mound of earth raised around a place for fortification or to hold out flood waters

Raze: to tear down to the ground

Reef: a rocky ridge or mound that extends up to the sunlit layers of the ocean, built by organisms (*i.e.*, coral)

Regional currents: movements of ocean water in a particular area (*i.e.*, near shore, near the equator)

Residence time (related to greenhouse gases): the length of time a gas stays in the atmosphere before it is altered or destroyed by chemical reactions, or physically removed from the environment

Respiration: any process used by organisms to generate energy (*i.e.*, use of oxygen and production of

carbon dioxide by animals to burn carbohydrates for energy)

Retention: capacity for holding in position

Salinity: the saltiness of a water solution

Sample: a small portion of a population considered to be representative of the whole for experimental purposes

Saturated solution: a solution that cannot dissolve any more of a substance

Sediment: particles derived from pre-existing rock, or biological or chemical processes, that are deposited on the Earth's surface or on the floor of a lake or ocean

Shoal: a sandy elevation of the bottom of a body of water, creating a shallow place

Silt: fine rock particles (smaller in grain-size than sand, but larger in grain-size than clay) often transported, and eventually deposited, by rivers and streams

Solstice: one of two times a year when the Earth is farthest from the Sun in its elliptical orbit

Spatial variation: a change in conditions with distance

Speleothem: a limestone spire that forms slowly as calcite dissolved in groundwater precipitates

Spires: a pointed formation or structure, as in a steeple or an icicle

Stalactite: a speleothem that hangs from the ceiling and tapers downward

Stalagmite: a speleothem that rises from the ground and tapers upward

Standard deviation: a statistical measure of the degree to which a set of values clusters about a mean value

Stomata: pores on plant leaves through which gases can enter or leave

Sublimation: process in which a substance changes state from a solid directly to a gas without passing through the intermediate liquid phase

Sulfur: element which forms many compounds (*i.e.*, SO_2 gas emitted by volcanoes that can form particles in the atmosphere that absorb heat)

Temperature gradient: change in temperature over distance (*i.e.*, altitude)

Temporal variation: a change in a property or in conditions over time

Terrarium: an enclosed container for growing or displaying plants

Terrestrial: growing, living or found on land, as opposed to in water; pertaining to Earth as distinct from other planets

Thorium: a radioactive element, some isotopes of which are produced by the radioactive decay of uranium

Topsoil: the uppermost layer of soil, rich in nutrients used by plants

Transpiration: evaporation of water through the leaves and stems of plants

Triatomic molecule: a molecule consisting of three atoms

Uranium: a radioactive element that decays over time to produce isotopes of thorium and lead

Water vapor: the gaseous state of water

Weather: atmospheric conditions in a given region over short time scales

Wetlands: areas of shallow water containing much vegetation

Annotated Bibliography

This annotated bibliography can also be used as a stand-alone piece for students and teachers studying global climate change. It offers descriptions of and excerpts from books that are interdisciplinary in nature, dealing with the relationship of climate change and the processes of biology, chemistry, physics, and human dimensions. Periodicals that cover global change topics are described and subscription information is provided.

The bibliography includes: **Books; Periodicals; Electronic Resources, Curricula, and Teacher-training Programs; and Related Topics**.

Books

Sierra Club Books
100 Bush Street
San Francisco, CA 94104
(415) 923-5600 (Bookstore)
(415) 291-1602 (fax)
WWW: http://
www.sierraclub.org/books/

The Coevolution of Climate and Life
Schneider, Stephen H., and Londer, Randi
(1984; ISBN#0-871-563-495/Out of Print)

This collaboration between a climatologist and science writer deals with the Earth's climate and its profound effect on life. While the content is complex, basic concepts are defined and specialized jargon is kept to a minimum, making the book accessible to most teachers and some older students. Part I is a good introduction to paleoclimate, taking the reader through "four billion years of weather." Part II includes an explanation of the climate system and what drives it, Earth's energy balance, computer climate modeling, and the causes of climate change. Part II focuses on the human dimensions—how climate affects us and how we might be affecting climate.

Yale University Press
P.O. Box 209040
New Haven, CT 06520
(203) 432-0940
(203) 432-0948 (fax)
WWW: http://www.yale.edu/yup/

The Changing Atmosphere
Firor, John
(1990; ISBN#0-300-056-648)

A very readable introduction to acid rain, global warming and ozone depletion by a scientist who clearly understands both the scientific research and its policy implications. Written in nontechnical language, this book should be accessible not only to teachers, but to many middle and high school students as well. Dr. Firor's treatment of complex scientific issues is unusually even-handed. At the same time, he does not skirt what he calls "the ultimate question: are we in trouble or not?" The book provides a good overview of the current state of atmospheric science, together with references to the scientific literature for teachers and students who wish to go further.

Clouds in a Glass of Beer: Simple Experiments in Atmospheric Physics
Bohren, Craig F.
(1987; ISBN#0-471-624-829)

Teachers and students will enjoy these practical demonstrations and laboratory experiments that bring a range of atmospheric phenomena into the classroom. Among the experiments described in this information-rich and entertaining volume are: Dew Drops on a Bathroom Mirror, On a Clear Day You Can't See Forever, The Green Flash, and Indoor Rainbows. More than a simple collection of experiments, this book is full of relevant material on how the atmosphere works. For each experiment, the author discusses the underlying physics, providing background information and explanations of basic concepts before describing the actual experiment. The explanations are thorough and by no means simplistic, making the book suitable primarily for teachers and advanced students willing to make a concerted effort.

John Wiley & Sons, Inc.
605 3rd Avenue
New York, NY 10158-0012
(212) 850-6000
(212) 850-6088 (fax)
WWW: http://www.wiley.com/

Atmospheric Change: An Earth System Perspective
(1993)
Graedel, T. E., and Crutzen, Paul J.

This comprehensive and authoritative book is full of clear information and illustrations on the subject of atmospheric change. While the book is technical in nature, and should be considered mainly as a teachers resource, many of the charts and graphs can be understood and used by students as well. Recognizing the highly interdisciplinary nature of the subject, the authors broaden the range of topics conventionally included in atmospheric science. Though some of the material is quite complex, those willing to make the effort will find much they can understand and use in the classroom. This extensive compendium of facts and figures serves as an excellent volume of reference for high school teachers and students exploring global change.

W.H. Freeman and Company
4419 West, 1980 South
Salt Lake City, UT 84104
(801) 973-4660
WWW: http://www-bookstore.ucsd.edu/
dev/BookSearch.html

State of the World
Brown, Lester, et al.
(1996; ISBN#0-393-038-513 hardcover; ISBN#0-393-313-395 paperback)

These annual reports from Worldwatch Institute's research team are a must for those who monitor and teach about the state of the planet. The series is probably the single most-used reference on current global environmental issues. The books are extremely well-researched and offer credible scholarship and up-to-date data. Each year, cutting edge topics are explored on a level accessible to teachers and high school students (and perhaps even some younger ones). The 1994 version contains chapters on carrying capacity, the forest economy, oceans, reshaping the power industry, transportation, computers, environmental health risks, cleaning up after the arms race, and the prospect of food insecurity.

W. W. Norton
National Book Company
800 Keystone Industrial Park
Granton, PA 18512
(212) 354-5500
(800) 458-6515 (fax)
WWW: http://web.wwnorton.com/

W. W. Norton
National Book Company
800 Keystone Industrial Park
Granton, PA 18512
(212) 354-5500
(800) 458-6515 (fax)
WWW: http://web.wwnorton.com/

National Science Teachers Association
1840 Wilson Blvd.
Arlington, VA 22201-3000
(800) 722-NSTA
(703) 522-6091 (fax)

University of California Press
1010 Westwood Boulevard
Suite 410
Los Angeles, CA 90024-2912
(310) 794-8158
(310) 794-8156 (fax)
WWW: http://
press-gopher.uchicago.edu:70/CGI/
AAUP/gais.cgi/1200/University%20of
%20Cal

Vital Signs

Worldwatch Institute

(1995; ISBN#0-393-037-819 hardcover; ISBN#0-393-312-798 paperback)

This new addition to the publications produced by Worldwatch is a concise companion volume to the State of the World reports. *Vital Signs* presents and analyzes a set of key indicators (44, in the 1994 issue) that track changes in environmental, economic, and social health. In text and graphics, the book provides current global data on trends in food, agriculture, energy, atmospheric constituents and temperature, economics, transportation, and a variety of other environmental and social matters. Most of the material is non-technical, making it accessible to students and teachers at many levels.

Project Earth Science: Meteorology

Smith, P. Sean, and Ford, Brent A.

(1994; ISBN#0-87355-123-0)

Project Earth Science: Meteorology is the second in the four-volume *Project Earth Science* series. The book is divided into three sections: activities, readings, and appendices. The activities are constructed around three basic concept divisions. First, students investigate the origin and composition of the Earth's atmosphere. Second, students examine some of the factors that contribute to weather. Third, students are introduced to concepts of air masses and the ways these masses interact to produce the weather around us. The readings which follow the activities enhance teacher preparation and introduce supplemental topics.

The Forgiving Air

Somerville, Richard C.J.

(1996; ISBN#0-520-088-905)

Perhaps the most accessible of all recent books on global change for lay-people is this volume from a meteorologist. As head of the Climate Research Division at Scripps Institution of Oceanography, University of California, San Diego, the author is a scientist who possesses both cutting-edge information and the ability to communicate this material to a non-technical audience. Written in entertaining and colloquial prose, this book is an excellent introduction for teachers and students into the range of issues known as global change, with a focus on climate change and ozone depletion.

Periodicals

Scientific American
(ISSN#0-036-8733)

This monthly magazine presents current scientific topics in a readable format for the layperson. Widely found in libraries and bookstores, *Scientific American* is accessible to teachers and students and often contains articles on global change topics. Recommended articles that focus on current topics in climate change include, "Core Questions: Glaciers and Oceans Reveal a Mercurial Climate," December 1993, and, "Sulfate Aerosols and Climatic Change," February 1994. Other significant global change articles include "Wetlands," January 1994, and "The Puzzle of Declining Amphibian Populations," April 1995.

Scientific American has published several excellent special issues which provide thorough overviews of their chosen subjects. Two issues worthy of note are: "Managing Planet Earth," September 1989, which contains 11 articles by leading authorities on a variety of global change topics, including especially good articles on atmospheric chemistry, climate change, water, biodiversity, and population; and "Energy for Planet Earth," September 1990, a collection of authoritative articles on solar, nuclear, and fossil fuel energy and their future prospects, along with articles on energy in the developing world, motor vehicles, and industrial energy use.

415 Madison Avenue
New York, NY 10017
(212) 754-0550
(212) 755-1976 (fax)
(800) 333-1199 (Subscriptions phone)
(515) 246-1020 (Subscriptions fax)
WWW: http://www.techexpo.com/toc/sci-am.html

New Scientist
(ISSN#0-262-4079)

This British weekly summarizes developments in science, including climate and other global change subjects, in a way that is entertaining, informative, and highly readable. It also has the advantage of being decidedly international in scope. *New Scientist* is a good resource for teachers and students who want more detail than is offered by a highly abbreviated publication such as *Science News*, but a far less technical presentation than the primary source journals such as *Science* and *Nature*. Each issue contains brief reports on current developments as well as lengthier review articles on important topics. Recent articles of relevance to climate change include, "Planes fly through climate loophole," January 5, 1995, "Greenhouse warming goes to market," January 21, 1995, "Rain moves north in global greenhouse," March 4, 1995, "Seeds of our own destruction," April 8, 1995, and "Price of life sends temperatures soaring," April 1, 1995.

New Scientist Subscriptions Department
IPC Magazines, Ltd.
King's Reach Tower
Stamford Street
London SE1 9LS
England
0-171-261-5000
0-171-261 6464 (fax)
WWW: http://www.newscientist.com

Science Service, Inc.
1719 N Street, NW
Washington, DC 20036
(202) 785-2255
(202) 659-0365 (fax)

Science News
(ISSN#0-036-8423)

A good secondary source for those seeking a condensed weekly update on a wide variety of scientific developments, including many issues relevant to the study of global change. The publication is brief, making it practical for those who already have too much to read but don't want to miss significant happenings in the sciences. The writing level is generally geared to an educated audience, but can occasionally get a bit more technical. Brief summaries of articles published in Science, Nature, and other primary-source journals are often included.

AAAS
1333 H Street, NW
Washington DC 20005
(202) 326-6666
(202) 842-5196 (fax)
WWW: http://science-mag.aaas.org/science/

Science
(ISSN#0-036-8075)

The premier publication in the United States for primary scientific research on a variety of topics including chemistry, physics, biology, geology, and medical sciences. Global change topics are covered in an in-depth manner. The research articles, reports, and technical comments are usually not geared to the layperson, but the sections devoted to news, commentary, perspectives, political developments, book reviews, etc., can be accessible and useful for teachers and advanced students. Science is published by the American Association for the Advancement of Science (AAAS).

National Science Teachers Association
1840 Wilson Blvd.
Arlington, VA 22201-3000
(703) 243-7100
WWW: http://www.nsta.org

The Science Teacher
(ISSN#0-036-855-5)

The Science Teacher is the professional journal for junior and senior high-school science teachers, offering articles on a wide range of scientific topics, innovative teaching ideas and experiments, and current research news. Also offers reviews, posters, information on free or inexpensive materials, and more. Published nine times a year, September through May.

Nature
345 Park Avenue South
New York, NY 10010
(212) 726-9200
(212) 696-9006 (fax)
WWW: http://www.nature.com/

Nature
(ISSN#0-028-0836)

The foremost British science journal, similar in scope and stature to Science. The technical level is high, making the majority of research articles inappropriate for a lay audience. However, like Science, there are many sections of the magazine that are far more accessible and sometimes help to explain the more technical research articles. The weekly issues contain editorial columns, reports on scientific news items including national and international political developments that affect science, correspondence, book reviews, and special features. One of most interesting and accessible features is the "News and Views" section which explains and elaborates on the research articles that appear later in the journal, or in other scientific publications.

Electronic Resources, Curricula, and Teacher-training Programs

Environmental issues are critically important agenda items in this country's geopolitical, economic, and social debates. From the halls of government to our individual homes, we are engaging in a nationwide search for balanced solutions to pressing problems. Scripps Institution of Oceanography (SIO) has existed since 1903 to study the converging needs of science and society. Its research programs embrace all the planetary processes in which our oceans play a part, including the biological, botanical, ecological, chemical, geological, meteorological, paleontological, and physical.

Current information about ongoing research at Scripps is offered via *SIO Explorations*, an engaging, beautifully illustrated periodical for the interested public. Published quarterly, it is available to Friends of the Aquarium through the Membership Department of the Stephen Birch Aquarium-Museum. Information about Scripps and its multitude of research, education, and other programs is also available from SIO's web site. To learn more about SIO's offerings for teachers and students, access the aquarium-museum's home page. Interested in services ranging from classes and recommendations on reading matter to an on-line question and answer service? Stay in touch!

SIO, UCSD, 0207
9500 Gilman Drive
La Jolla, CA 92093-0207
(619) 534-4109
(619) 534-6692 fax
WWW: http://sio.ucsd.edu/res_groups

SIO Aquarium-Museum
WWW: http://aqua.ucsd.edu

A new set of on-line resources developed by the Aspen Global Change Institute (AGCI) is now available from the U.S. Global Change Research Information Office (GCRIO). A key benefit of these resources is that they are continuously updated, providing current material in a rapidly changing field. All of the resources described below can be found at the AGCI World Wide Web address.

AGCI
WWW: http://www.gcrio.org/
agci-home.html

EarthPulse Notes: Searchable Summaries of Current Global Change Research is AGCI's database of article summaries from *Science, Nature, New Scientist, Scientific American, EOS,* and occasionally other publications.

WWW: http://www.gcrio.org/
ASPEN/
epnews/epnotes.html

EarthPulse NEWS: An Educator's Global Change Update is designed to provide K-12 teachers and their students with an accessible summary of the latest developments in the field of global change, with an emphasis on climate. EarthPulse NEWS is sponsored by the Stephen Birch Aquarium-Museum and the Center for Clouds, Chemistry and Climate, both of the Scripps Institution of Oceanography, University of California, San Diego.

WWW: http://www.gcrio.org/ASPEN/science/
EPNews.html

The EnviroLink Network
WWW: http://www.envirolink.org/

The Environmental Education Network
WWW: http://www.envirolink.org/
enviroed/

ERIC Clearinghouse for Science,
Mathematics, and Environmental
Education
WWW: http://www.ericse.ohio-
state.edu/

WWW Virtual Library: Environment
WWW: http://ecosys.drdr.virginia.edu/
Environment.html

Virtual Libraries: A number of World Wide Web home pages include comprehensive lists of environmental organizations and resources available on the Internet. They include links to numerous home pages operated by government agencies, educational institutions, companies, and non-profit organizations. Many of them have educational resources that can be down-loaded.

Classroom Earth/CIESIN
User Services Department
2250 Pierce Road
University Center, MI 48710
(517) 797-2700
(517) 797-2622 (fax)
WWW: http://naic.nasa.gov/naic/guide/
templates/class-earth-bbs.html/

Classroom Earth BBS. Consortium for International Earth Science Information Network (CIESIN) and Saginaw Valley State University, Saginaw Valley, Michigan. Lesson plans on global systems delivered through the Internet.

304 Recitation Hall
WestChester University
WestChester, PA 19383
(610) 436-2393
(610) 436-3102 (fax)

Educational Center for Earth Observation Systems. West Chester University, West Chester, PA. Instructional material and documents.

UCAR
P.O.Box 3000
Boulder, CO 80307-3000
(303) 497-1000
(303) 497-8610 (fax)
WWW: http://home.ucar.edu/

Global Change Instructional Program. UCAR, University Center for Atmospheric Research, Boulder, CO. Fact sheets, Macintosh diskettes, and slide set for the classroom.

West 79th Street
Central Park West
New York, NY 10024-5192
(212) 769-5100
(212) 769-5329 (fax)

Global Warming: Understanding the Forecast. Department of Education, American Museum of Natural History, New York, NY. Print materials on greenhouse gases and global warming.

Lawrence Hall of Science
University of California
Berkeley, CA 94720
(510) 642-5132
(510) 642-1055 (fax)
WWW: http://www.lhs.berkeley.edu/
lhshome.html

Global Warming and the Greenhouse Effect. Lawrence Hall of Science, University of California, Berkeley. Teaching guides for junior and high school teachers.

119 National Center
Reston, VA 22092
(703) 648-4460
(703) 648-4466 (fax)
WWW: http://www.usgs.gov/

GeoMedia2 Educational System. U.S. Geological Survey (USGS), Reston, VA. Multimedia and Hypermedia textual question answer stacks on global environmental change.

Global Change Information Packet. National Agricultural Library (NAL), U.S. Department of Agriculture. This bibliography has extensive information about global change issues, including lists of books and journal articles. The NAL is the world's largest agricultural library.

Reference Section
Room 111
National Agricultural Library
10301 Baltimore Blvd.
Beltsville, MD 20705

Global Environmental Change Interviews With the Experts: "Atmospheric Chemistry Causes and Effects." Special Assistant for Global Change, URI Graduate School of Oceanography, University of Rhode Island, Narragansett. Videos on selected topics.

South Ferry Road
Narragansett, RI 02882-1197
(401) 874-6222
(401) 874-6889 (fax)
WWW: http://www.gso.uri.edu/

The Heat is On: From the Greenhouse Effect to the Hole in the Ozone. Pacific Environment and Resource Center, Sausalito, CA. Curriculum units for the classroom.

Fort Cronkhite
Building 1055
Sausalito, CA 94965
(415) 332-8200
(415) 332-8167 (fax)

LEARN. Project Learn, NCAR, Boulder, CO. Fact sheets and print curricula on the atmospheric sciences and global climate change.

UCAR
P.O.Box 3000
Boulder, CO 80307-3000
(303) 497-1000
(303) 497-8610 (fax)
WWW: http://www.ucar.edu/
homepage.html

Project Atmosphere. American Meteorological Society, Washington, D.C. A comprehensive training program designed to enhance teacher effectiveness in generating interest and understanding in science, technology and mathematics among pre-college students.

Project Atmosphere
American Meteorological Society
1701 K. St. NW
Suite 300
Washington, DC 20006
(202) 466-6070
(202) 466-5729 (fax)
WWW: http://atm.geo.nsf.gov/11/AMS/

Tropical Ocean-Global Atmosphere Project (TOGA). Physical Oceanography Distributed Active Archive Center, Jet Propulsion Laboratories, Pasadena, CA. CD-ROM data sets of climate and weather observations from 1985-1990.

4800 Oakgrove Drive
Pasadena, CA 91109
(818) 354-4321
(818) 354-3437 (fax)
WWW: http://www.cms.udel.edu/coare/
toga.html

Reports to the Nation Series. OIES/NOAA, UCAR, Boulder, CO. Print materials and videos on varied topics (*i.e.,* El Niño and climate prediction).

UCAR
P.O.Box 3000
Boulder, CO 80307-3000
(303) 497-1000
(303) 497-8610 (fax)
WWW: http://www.ucar.edu/
homepage.html

1709 New York Avenue, NW
Suite 700
Washington, DC 20006
(202) 638-6300
(202) 638-0036 (fax)

Teacher's Guide to World Resources. World Resources Institute, Washington, D.C. Climate and global change narrative and strategies for action.

Related Topics

ICESS
6th floor
Ellison Hall
University of California
Santa Barbara, CA 93106
(803) 893-8095
(803) 893-2578 (fax)

An El Niño Scenario. Earth-Space Research Group. University of California at Santa Barbara. Hypermedia encyclopedia with text and graphics available on Mosaic.

2801 South University
Little Rock, AR 72204
(501) 569-3113
(501) 569-8694 (fax)

Arkansas' Project MAST: Mathematics and Science Together. College of Education Gifted Programs, University of Arkansas, Little Rock. Print materials integrating meteorologically-oriented science and mathematics instruction.

The Weather Underground
University of Michigan
1541 Space Research Building
Ann Arbor, MI 48109-2143
(800) 386-4141
(313) 936-0503
email: blueskies@umich.edu
WWW: http://
groundhog.sprl.umich.edu

Blue Skies. The Department of Atmospheric, Oceanic and Space Sciences, University of Michigan, Ann Arbor. Real-time meteorological data sets and instructional materials for classroom use.

University of Hawaii
1776 University Avenue
Honolulu, HI 96882
(808) 956-7703
(808) 956-3106 (fax)

Developmental Approaches in Science and Health (DASH). College of Education, University of Hawaii-Manoa, Honolulu. Print materials and teacher training on a wide range of science-education topics.

1155 16th Street, NW
Washington, DC 20036-4800
(202) 872-4600
(202) 872-4615 (fax)

Foundations and Challenges to Encourage Technology-Based Science (FACETS). Module title: Under the Weather. Education Division of the American Chemical Society, Washington, D.C. Student materials and a newsletter.

Lawrence F. Lowery
FOSS
University of California Berkeley
Berkeley, CA 94720
(415) 642-6000

Full Option Science System (FOSS). Encyclopedia Educational Corporation, University of California, Berkeley. Student equipment, print materials, and alternative assessment materials on varied topics.

222 North 20th Street
Philadelphia, PA 19103
(215) 448-1200
(215) 448-1235 (fax)

Franklin Activity Kits. Franklin Institute Science Museum Staff, Philadelphia, PA. Science activity kits, print materials and teacher training for hands-on learning experiences in Earth sciences.

Improving Middle School Science: A Collaborative Approach. Education Development Center, Newton, MA. Print materials for 7th and 8th grade students integrate life, physical, and Earth sciences.

55 Chapel Street
Newton, MA 02158
(617) 969-7100
(617) 969-3401 (fax)

Insights: A Hands-on Elementary Science Curriculum. Education Development Center, Newton, MA. Earth science with teachers' guides, and kits of science materials.

55 Chapel Street
Newton, MA 02158
(617) 969-7100
(617) 969-3401 (fax)

Joint Education Initiative. USGS and University of Maryland, University of Maryland, College Park. CD-ROM data sets for use in teaching geology and other Earth sciences.

Department of Geology
University of Maryland
College Park, MD 20742
(301) 405-4090
email: staff@jei.umd.edu

Kids as Global Scientists. National Science Foundation, University of Colorado, Boulder. Internet connectivity of school-based laboratories for collection and interpretation of weather phenomena.

University of Colorado Boulder
Campus Box B 19
Boulder, CO 80309-0019
(303) 492-0111
email: songer@stripe.colorado.edu

National Geographic Kids Network "Weather in Action." Technical Education Research Center, Cambridge, MA. Kits including software and sample files, teacher's guide, software manual, wall maps, materials, on-line subscription, and student handbooks.

2067 Massachusets Avenue
Cambridge, MA 02140
(617) 547-0430
(617) 349-3535 (fax)

Project First Step (Science and Technology Education Program). U.S. Space Foundation, Colorado Springs, CO. Print materials using space and aviation technologies to teach Earth and physical sciences.

2860 South Circle
Suite 2301
Colorado Springs, CO 80906-4184
(719) 576-8000
(719) 576-8801 (fax)

Renewables are Ready: A Guide to Teaching Renewable Energy in Junior and Senior High School Classrooms. Union of Concerned Scientists, Cambridge, MA.

2 Brattle Square
Cambridge, MA 02238-9105
(617) 547-5552
(617) 864-9405 (fax)

Terracorps. Upland Unified School District, Upland, CA. Instructional modules with videotape on Earth science topics.

390 North Euclid
Upland, CA 91786
(909) 985-1864
(909) 949-7862 (fax)

Metric Conversion Table

Government and industry in the U.S. have largely converted to metric system, although English units can still be found along with the metric on many commercial products. The U.S. is the only country in the world whose citizens don't primarily use metric—especially surprising because Thomas Jefferson was one of the originators of the metric system in the 1790s. In the 1820s, John Adams recommended U.S. adoption of metric as an international standard; in the 1890s, the meter and international kilogram became the legal U.S. standards for length and mass. Government and industry conversion was adopted in 1992. Until U.S. citizens finally give up the old system, however, converting from English to metric units or vice versa will sometimes be necessary. The following table shows how to do the most common conversions.

Rules for Metric to English Conversion

Multiply metric	by	this number	to get	English
___ centimeters	x	.394	=	___ inches
___ meters	x	3.281	=	___ feet
___ kilometers	x	.621	=	___ miles
___ liters	x	1.057	=	___ quarts
___ liters	x	.264	=	___ gallons
___ kilograms	x	2.205	=	___ pounds
___ °C	x	1.8 + 32	=	___ °F

About the Photographs

The photographs on the front and back covers show a variety of Earth's climatic zones.

Front Cover

Counterclockwise, beginning in the upper left:

Algal bloom in the Pacific Ocean Temperature variations affect the rate of plant growth in the oceans. Photo by Tracy Weir.

Desert sand Temperature changes affect local or global water cycles, causing some regions to receive more water and others to receive far less. Photo by Susan R. Green.

Temperate climate farmland Agriculture type and distribution can affect the amount of heat-trapping gases in the atmosphere. Photo by Susan R. Green.

Mt. Fuergo, Guatemala Volcanic eruptions such as this one emit heat-trapping gases into the atmosphere. Photo compliments of National Center for Atmospheric Research.

Palm trees in the tropics Trees, like other plants, help regulate atmospheric content by absorbing carbon dioxide and other gases, and by producing oxygen. Photo by Heidi J. Hahn.

Photographs of students were taken by Monica C. Hamolsky, Ann Scott Parker, and Deborah Zmarzly.

Back Cover

Quelccaya ice cap, southern Peruvian Andes This photograph of an ice cap at an elevation of 5,670 meters, high in the Andes, was taken in 1977. The 50-meter-thick ice cap showed annual layers of snow separated by dry-season dust, representing 1,500 years of climate change. By August, 1995, the entire ice cap had disappeared as a result of increasing temperature. Photograph by Lonnie Thompson, Byrd Polar Research Center.